Smart Design

Philip Breedon
Editor

Smart Design

First International Conference Proceedings

Springer

Philip Breedon
Nottingham Trent University
Burton Street
Nottingham NG1 4BU, UK

ISBN 978-1-4471-2974-5 e-ISBN 978-1-4471-2975-2
DOI 10.1007/978-1-4471-2975-2
Springer London Heidelberg New York Dordrecht

British Library Cataloguing in Publication Data
A catalogue record for this book is available from the British Library

Library of Congress Control Number: 2012936117

Printed on acid-free paper

Springer is part of Springer Science+Business Media (www.springer.com)

Programme Committee

Dr. Ric Allot, Director, Photonics and Plastic Electronics, KTN, UK
Professor Stephen Barrass, University of Canberra, Australia
Professor David Brown, NTU, UK
Dr. Luigina Ciolfi, Course Director, Digital Media Design Programme, Computer Science and Information Systems, University of Limerick, Ireland
Professor Ann Cranny Francis, Professor of Cultural Studies & Director, Transforming Cultures Research Centre, University of Technology, Sydney, Australia
Professor Tilak Dias, NTU, UK
Dr. Kelly Dobson, Department Head of Digital + Media, Rhode Island School of Design, Providence, Rhode Island, USA
Dr. Michael Hohl, University of Huddersfield, UK
Professor Jocelyn Faubert, Université de Montréal School of Optometry, NSERC-Essilor Chair on Presbyopia and Visual Perception, Canada
Professor Steve Goodhew, Associate Dean of Research, NTU, UK
Professor Moyez Jiwa, Professor Health Innovation, Chronic Disease Curtin Health Innovation Research Institute & Chair Curtin Health Innovation Research Institute, Western Australia
Professor Jim Johnston, Professor of Chemistry, Victoria University of Wellington, New Zealand
Dr. Sarah Kettley, NTU, UK
Dr. Ahmad Lofti, NTU, UK
Dr. John Marshall, University of Michigan, USA
Dr. Mark Norman, Merlin Robotic Systems, UK
Dr. Johan Redström, Studio Director, Interactive Institute's Design Research Unit. Göteborg, Sweden
Dr. Mariana Salgado, User-Driven Innovation Centre, Laurea University of Applied Sciences, Helsinki, Finland
Dr. Holger Schnädelbach, Mixed Reality Laboratory, Nottingham University, UK
Professor Bob Stevens, Rutherford Appleton Laboratory, UK

Dr. Louise Valentine, Duncan of Jordanstone College of Art & Design, Dundee, UK
Clinical Professor Michael Vloeberghs, Nottingham University Hospital, UK
Dr. Tom Waller, Aqualab, Speedo, UK
Dr. Brian Weeks—IPI Europe, UK
Dr. Hazel White, Programme Director, Master of Design, Duncan of Jordanstone College of Art and Design University of Dundee, UK
Professor Alan Winfield, EPSRC Senior Media Fellow, Director of Science Communication Unit, University of the West of England, Bristol, UK

Foreword

The introduction of new materials and rapidly changing technologies is assigned huge and unprecedented importance in the economy of the twenty-first century. Good product designs merge materials, technology and hardware into a unified user experience, where the technology recedes into the background and people benefit from the capabilities and experiences available. By focusing on functional gain, critical awareness and emotive connection, even the most multifaceted and complex technology can be made to feel straightforward and become an integral part of daily life.

Designers and developers must understand how to progress and harness knowledge to inform their innovations. They need to know how to explore the opportunities in order to become design innovators and autonomous problem solvers, combining their practical skills with an awareness of aesthetics, practical applications and function.

Design students at Nottingham Trent University develop their Smart design skills and knowledge based on a platform of considered research, and on an interdisciplinary approach to learning. They gain the ability to look for requirements and opportunities and respond to them by developing a range of ideas and designing innovative products and systems.

I hope this first International Conference on Smart Design will inspire and motivate you. We aim to provide an opportunity for you to contribute to and gain a broader understanding of design, materials and technology based research; ultimately allowing you to develop concepts and ideas allied to the cohesion of new and emerging technologies within your own designs.

Philip Breedon

Acknowledgments

The first International Conference on Smart Design would not have been possible without the support, hard work and endless efforts of a large number of individuals.

A large number of individuals and support groups have been instrumental in facilitating the conference and this publication, in particular, Professor Wayne Cranton and Dr. Rob Morris from the School of Science and Technology; Melanie Bentham-Hill Research and Graduate office and Jo Frith, Management Accountant; Jo Hall, Marketing Manager, Ed Pearson, Marketing Executive and James Lazenby Events Manager, all at Nottingham Trent University.

Special thanks for their continual support goes to Ann Priest, Pro Vice Chancellor and Head of College for Art, Design and the Built Environment and Peter Westland, Dean of Architecture, Design and Built Environment.

Sincere gratitude goes to our sponsor, PepsiCo Europe and our partner the Australasian Medical Journal who supported our ideas and gave advice.

Special gratitude goes to Matthew Malpass for his invaluable advice, help and support in compiling this publication, and to David King for all his efforts in relation to coordinating this publication and his help with many aspects of conference organisation.

Contents

Chapter 1
Rethinking Packaging: How Electronic Packs for Pharmaceuticals Work with Mobile IT to Improve Patient Adherence to Medication Regimens

James Woudhuysen and Peter Rivers

Abstract Indictments of waste, and distaste for the packaging of consumer goods, are widespread in modern life. Yet in England, at least, households account for less than a third of overall waste, and household waste—like the industrial and commercial kinds—is actually in decline. In addition, the potentialities of packaging as a force for progress are too often ignored. Electronic packaging could improve the way users of prescription drugs take their medicines. With an ageing population taking more and more varied kinds of drugs, patient adherence to medication regimens is a growing social and economic issue. Linked to mobile IT, electronic packaging can, in principle, do much to ensure that medical patients wind up taking the right drugs at the right times. The paper reviews four examples of electronic packaging, and goes on to situate them in the context of mobile health (mHealth)—mobile telephony applications that prompt patients, each time they take medicine, to report their symptoms very simply and, if necessary, receive advice back from doctors. Together, reporting adherence and symptoms can enable prescribers to build up a rich picture of the effectiveness of each prescription issued. Doses or medication can be changed, or medication withdrawn, using real intelligence about patients. The dual approach outlined here can also reduce the problems that arise with the use of multiple medications by a patient. Innovation in the way in which prescription drugs are packaged could do a lot for society. This paper explores the potential of linking the electronic packaging of medicines to mobile IT, as well as the attitudinal and likely regulatory barriers that could impede progress in this domain.

J. Woudhuysen (✉) · P. Rivers
De Montfort University, Leicester, UK
e-mail: james@woudhuysen.com

P. Rivers
e-mail: privers@dmu.ac.uk

P. Breedon (ed.), *Smart Design*,
DOI: 10.1007/978-1-4471-2975-2_1, © Springer-Verlag London 2012

Keywords Waste · Health · Health economics · Ageing · Polypharmacy · IT · Applications · Patient-centric · Pharmacovigilance · Security · Privacy · Managed outcomes · Regulation · Consumer goods

1.1 Introduction: Packaging as Waste—or as Progress

The issue of waste—and, within that, packaging waste—is very prominent in Britain today. In late June 2011, Clean Up Britain, a national anti-litter campaign, was launched, 'backed by more than 40 leading sports stars, television and showbiz celebrities' [1]. Earlier that month, UK environment secretary Caroline Spelman was reported to be planning to extend a voluntary code on reductions of packaging from food to children's toys, and perhaps on to mobile phones and electrical goods like microwaves and toasters. The same report noted that Britain generates 29 Megatonnes (Mt) of household waste every year, of which more than 20%—6 Mt—is packaging; and that, of this 6 Mt, 30% is estimated to come from toys, mobile phones, computers and other appliances [2].

The previous year, in Phase 2 of the so-called Courtauld Commitment, 29 major retailers and brand owners volunteered, by 2012, to

- Cut the weight, increase recycling rates and increase the recycled content of all grocery packaging, with the aim of cutting its carbon impact by 10%
- Reduce UK household food and drink waste by 4%
- Reduce traditional grocery product and packaging waste in the grocery supply chain by 5%—including both solid and liquid wastes [3].

Britain is not alone in its hatred for and guilt about packaging. In the Design For Good blog, run on America's prestigious knowledge forum bigthink.com, one commentator says this about 'rethinking' packaging:

> Product packaging is one of consumerism's most toxic byproducts—transient, temporary, and lacking the vaguely utilitarian excuse for existence that the product it contains can claim. It requires energy to make, adds to shipping weight, and is often made of polymers that linger in landfills for thousands of years. Now, designers are turning to innovative materials and engineering to revolutionize the environmental impact of packaging [4].

With the claim that packaging is 'most toxic', the writer continues in the long line of all the books that now identify business leaders, parents, children, comics, dentists, body fat, friends and much else besides as toxic, too.

Today most discussions on packaging focus on the need to minimise it.[1] This paper gives a hint, by contrast, of *what packaging might do*—apart from try more or less to abolish itself. In the case of pharmaceuticals, clever packaging can be a force for progress.

---

[1] Discussions on packaging parallel those on energy, which typically focus on the need to minimise energy use, and not on the need to innovate in energy supply [5].

The broadly conservationist approach to packaging is open to challenge. First, in England in 2009, industrial sector waste amounted to 24.2 Mt and commercial sector waste 23.8 Mt [6]. Between the financial year 2009/2010 and the year July 2009–June 2010, by contrast, the amount of waste collected from household sources in England was lower than each of these sectors: 23.6 Mt [7]. Those who attack packaging usually condemn it as one of the excesses of consumerism. They forget that in England, at least, the consumer dimension of waste is less than a third of the country's overall tonnage.

Second, with all three kinds of waste, there has been a *reduction* in tonnage: as the number of firms in UK industry has fallen by more than 18% since 2002/2003, so industrial wastes have dropped by 13.4 Mt, or 36%. Yet as the number of firms in UK commerce rose by 12% over the same period, the waste generated by commerce has fallen by 6.5 Mt, or 21%. As for the household sector, over the 5 years to 2009/2010 there was an average yearly decrease in the weight of waste of 1.6%, even if annual declines only emerged after 2006/2007. The decline in household waste is the more remarkable given that the number of people living in England is estimated to have risen from 50.11 million in mid 2004 to 51.81 million in mid 2009 [8]. Clearly the UK is already generating less waste per householder than it did 5 years ago.

Third, the *functionalities of packaging in use* are ignored whenever it is viewed simply as waste. It is true that media around the packaging industry cover how packs preserve food, drink, toys and other products from wear and tear, health hazards and many other problems. Yet when they discuss *how to put electronics to work in packs*, packaging industry media mainly cover the electronic tagging of products for supply chain management and for security; and though such functionalities are important to corporations, they are less so to end-users.

In the case of the packaging of pharmaceuticals, by contrast, the benefits of packs to end-users are much clearer. For decades the contraceptive pill has been available in calendar packs that tell the user when to take it. Now, however, electronic packaging can improve the efficacy of treatments, and also cut costs for those who hold budgets for healthcare. To understand why, a short excursion is needed into what is known as patient *adherence* to medication regimens.[2]

1.2 Why Adherence to Medication Regimens is a Vital Issue

Though it is hard to establish exactly when he said it, US Surgeon General C Everett Koop once famously remarked: 'Drugs don't work in people who don't take them.' In fact as far back as 1997, the cost to America of illness due to non-adherence was estimated at $100 billion a year [10]. In Britain the cost of non-adherence in terms of new admissions to hospitals may not, today, much

[2] Some debate accompanies the medical categories adherence to, compliance with and concordance with medication regimens [9].

exceed £200 million [11]. However, with the ageing of the population, more patients will tend to take a variety of drugs during the course of a day ('polypharmacy'), a situation that is likely to raise the costs of non-adherence.

Poor adherence to medication regimens can lead to outdated products staying in medicine cabinets, recurrent illness, and poor data in clinical trials. It has been argued that research aimed at reducing non-adherence may have more impact on population health than the development of new drugs [12]. Non-adherence may account for up to 80% increased risk of death in diabetes [13], and a 3.8 fold increased risk of death in the year following a heart attack [14].

Historically, responsibility for adherence has rested with the patient or carer. Although calendar contraceptive packs showed the pharmaceutical sector's sensitivity to adherence issues, medical packaging is primarily designed for convenience in storage, portability, and maximum chemical longevity of the drug.

Despite these longstanding limitations, the world market for pharmaceutical packaging, valued at $42 billion in 2008, is forecast to reach $68 billion by 2015 [15]. Clearly, packaging is the interface between patients and something even bigger: the pharmaceutical sector. Many have long stigmatised that sector [16]. Yet critics of both packaging and of Big Pharma must now deal with some undeniable facts:

- Electronic containers of pharmaceuticals are already on the market
- These kinds of 'smart' packs can link up to a growing number of health applications carried on mobile phones, or on tablet computers. Together, software and hardware in this field are generally abbreviated as mHealth[3]
- Together, electronic packs for drugs and mHealth applications may help ensure that patients actually take all the drugs they are prescribed at the right time and in the right dosage.

1.3 Electronic Packaging as a Means of Measuring Adherence

In the late 1980s, electronic methods of recording patient behaviour automatically, in real time, emerged in the form of Medication Events Monitoring Systems (MEMS). MEMS are plastic containers into which medicine is placed. Whenever a pack closure is removed or a dose pushed out from a blister pack, in-pack electronics enable the exact time of the event to be recorded. Events are usually plotted as a pattern of dots on a timeline running over a number of months. It is the overall patterns, not the individual dots, which are revealing.

MEMS containers were usually only used in clinical trials, or in experiments—to test their feasibility [17, 18]. With advances in IT, MEMS can be built into

[3] mHealth is a close relative of eHealth, or health care conducted with the aid of fixed-line telecommunications.

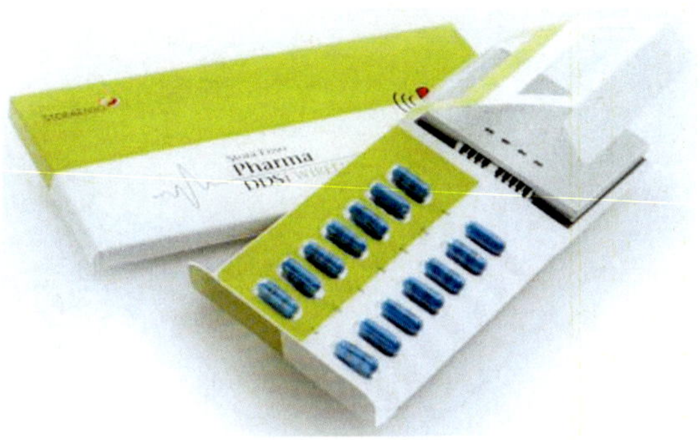

Fig. 1.1 Stora Enso's Pharma Discrete Dosage Slider Intelligent Wireless

blister packs, and mHealth can allow MEMS to work alongside text entry, by the patient, of data that describe his or her symptoms. Below, we list four examples of electronic packaging for adherence:

1. Headquartered in Helsinki and Stockholm, Stora Enso, a €10 billion company specialising in packaging, paper and wood products [19], makes an electronic pack named Pharma Discrete Dosage Slider intelligent Wireless (DDSi WL). Here the carton board upon which blisters are based is coated with conductive ink. On the board also is an RFID microchip, made by Sweden's Cypak and powered by a rechargeable battery for which each charge is good for several months. As each pill is removed, Near Field Communications transmit data wirelessly to the health records of a doctor, pharmacist, healthcare provider, or relative. A beeper can sound to remind the patient to take a pill, and a mobile phone can be used to display data, keep a diary, or send an SMS reminder to take the drug [20] (Fig. 1.1).
2. Qolpac, a company based in the Netherlands and Germany, offers what it calls Objective therapy Compliance Measurement (OtCM). Here the designs are similar to that of Stora Enso's product, with the difference that conductive tracks and RFID tags are applied as self-adhesive labels to standard blister packs [21] (Fig. 1.2).
3. MeadWestvaco (MWV) Corporation is based in Richmond, Virginia, USA. Its Cerepak® blisterpacks are based on the conductive/RFID electronic technologies already outlined [22] (Fig. 1.3).
4. MWV also makes Helping Hand, designed by Designit, and the winner of several prizes for its design [23]. Customised for particular blister packs and particular medication schedules, this curved plastics holder flashes a light and sounds an alarm when a pill is due to be taken. The patient pulls out the blister pack, takes the pill and reinserts the pack into its holster. The holster registers what has happened, and, through 'traffic light' signals, gives a visual indication to tell patients how they good their adherence is. A data capture version allows patient information to be uploaded though a USB cable (Fig. 1.4).

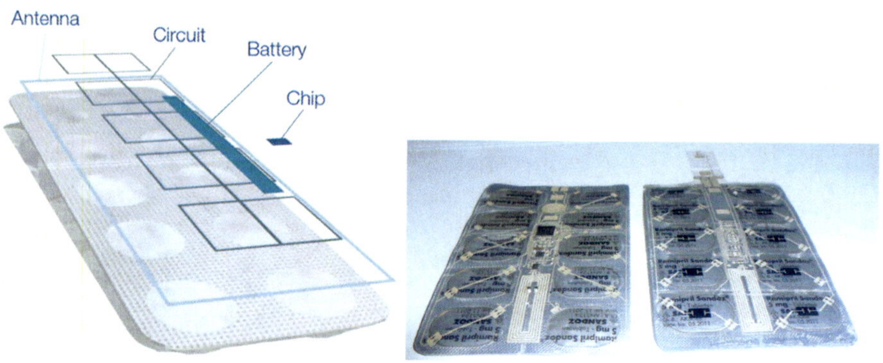

Fig. 1.2 Principles behind the electronic packs of Qolpac, and their application to Sandoz's Ramipril tablets

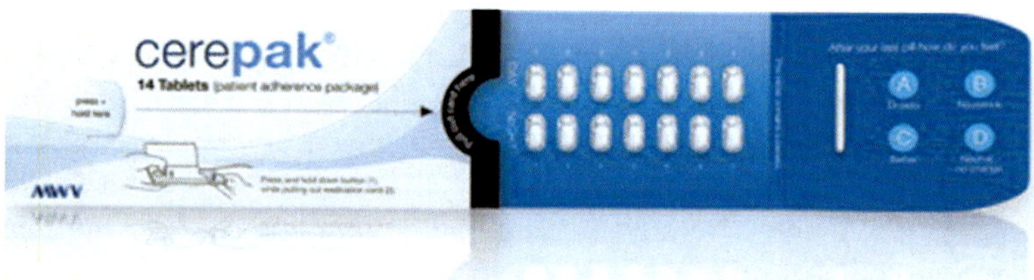

Fig. 1.3 MWV's Cerepak®

Fig. 1.4 MWV's Helping Hand

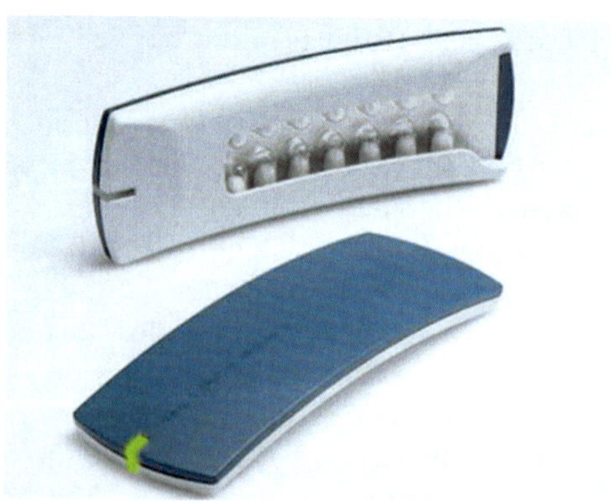

1.4 Pharmaceutical Packaging in the Context of mHealth

All the above packs can work with mobile phones, even if some also link up to dedicated electronic readers of the information they supply. Together, the automatic reporting of adherence patterns and the recording by patients of symptoms enable prescribers to build up a rich picture of the effectiveness of each prescription issued. Doses or medication can be changed, or medication withdrawn, using real intelligence about patients. Already, Vodafone offers what it calls Mobile Patient Reported Outcomes (mPRO) software to healthcare specialists. That software allows patients easily to report the level of pain they are feeling, their body temperature, or the kind of exercise they have done on the day of reporting [24].

Vodafone's focus on outcomes and on exercise is telling. In what has become known as a *patient-centric* approach to medicine (one that orientates to the patient's full range of needs), drug and biotechnology manufacturers now show a lot of interest in *all* aspects of patient treatment. As PricewaterhouseCoopers puts it, in future, drug companies

> will have to switch from selling medicines to managing outcomes. They'll have to bundle different products together and supplement their therapies with health management services like compliance monitoring, dietary guidance and fitness regimes. However, most companies won't be able to create packages of branded medicines and generics for different conditions singlehandedly, so they'll have to collaborate with rival manufacturers. And few, if any, companies will be able to deliver all the services patients need, so they'll have to collaborate with numerous other organisations, including hospitals, clinics, technology vendors and lifestyle service providers [25].

In this world of managed outcomes, packaging has a more active role to play than just carrying drugs. Indeed according to Chris Penfold, a designer active in the field, the boundary between pharmaceutical packaging and *medical devices*

> …is becoming ever more blurred as the following areas converge: smartphone technology, materials technology, internet capabilities and bandwidth, Wi-Fi and Near Field Communications (NFC). This is particularly evident in the area of patient compliance (adherence) [26].[4]

Clearly much can be done around pharmaceuticals, packaging, medical devices and IT.

---

[4] In fact matters go further. In the US, three engineers and a biochemist look forward to drugs that contain RFID tags, so that their safe progress from factory to gastro-intestinal tract can be monitored. Their aim is 'an ingestible, biocompatible RFID-tagged pharmaceutical and biosensor (BRTP) to confirm ingestion by the patient, to improve the non-invasive collection of data, and to monitor internal body conditions, such as pH, temperature, and, perhaps one day, the amount of drug released into the patient's system'. Here electronic packaging would be not be necessary, since a 13.56 MHz bottle/cap reader would scan and record the number of RFID enabled pills within the patient's pill bottle, and a wand would scan the patient's abdominal wall to confirm that a specific pill had been ingested. The tag itself would be $2.5 \times 1.3\ mm^2$ and would be clothed in an impermeable, biocompatible plastic membrane [27].

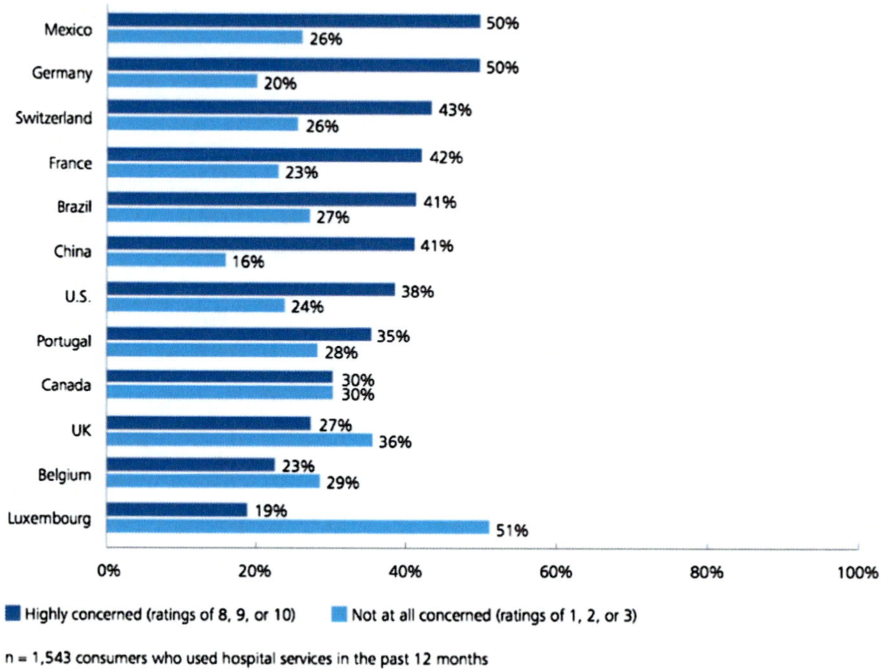

Fig. 1.5 Levels of concern about privacy and security around sharing personal health information with a doctor over the Internet, by country, 2011

For all the possibilities, however, the European Union (EU) does not seem yet to be exercised either by electronic packaging and mHealth. There is nothing on electronic packaging in EU legislation, and only one recent reference—unrelated either to packaging or to adherence—to mobile health [28]. It is a very similar story with the EU's Directorate for Consumers & Health: neither reference documents on EU legislation (EudraLex) for public health, nor the Directorate's Key Documents and Projects web pages on eHealth, refer to packs, adherence, or mHealth [29]. New EU regulations on 'pharmacovigilance', which are aimed at improving the safety and the benefits of drugs, are equally neglectful [30]. And when two EU commissioners spoke out in 2011 on eHealth, they too made no mention of packs, adherence, or mHealth [31].

How long this state of affairs can persist, however, is an open question. Despite the benefits of electronic packaging and mHealth, fears about personal privacy and security might well overwhelm what can properly be regarded as a rather modest pairing of two relatively simple technologies. In some countries—though not, it appears, in the UK—such fears may be quite widespread. According to the Deloitte Center for Health Solutions, when consumers were asked how concerned they are that the privacy and security of their medical information might be at risk if shared with doctors over the Internet, international responses were quite negative [32] (Fig. 1.5). Add these fears to the one-sided distaste felt for packaging and the suspicion that attaches to Big Pharma, and forward movement may prove rather difficult.

With mobile phones, continual reports 'failing to rule out' damage to the brain from radiation have not managed to dent popular enthusiasm for handsets. However, with electronic packaging and mHealth the dangers of 'bad' business practice, whilst slight compared with the benefits of the technologies, could inflame critical opinion. As a result, hasty and over-stringent regulation could still, in future, form a major barrier to their development.

1.5 Conclusion: Toward a Genuine Rethink of Packaging

The electronic packaging of drugs is, to some extent, a special case. Drugs are dear; so to add a proportionately modest increment to the cost of drugs by making their containers electronic, to improve efficacy, makes sense. It may be a decade or more before the packaging of mainstream consumer goods, which are generally much cheaper than drugs, begins to receive the electronic treatment.

Even in drugs, progress to date has not been rapid, and electronic methods are not yet the 'gold standard' for measuring adherence [33]. Nevertheless, the price of electronic packs is coming down. Applications in mHealth are multiplying.[5] Trends in health economics and demographics favour greater attention to adherence. Also, a growing body of regulation has made the long, small print in the instructions that accompany medicines almost impossible to read in their lengthy and miniaturised entirety; patients, therefore, might more easily track problems with their own drug regimens by branched closed questions on a mobile phone.

As Deloitte reports, younger adults have relatively carefree attitudes toward medical IT [35]. On the other hand, older people urgently need what electronic packaging and mHealth can bring them. For how much longer will complacent attitudes to adherence, and the impending threat of regulation, combine to hold back the breakthroughs that are so evidently possible, and so evidently necessary?

Whatever the answer, it should be clear that the role of packaging, both in and beyond pharmaceuticals, is far from being completely negative. It is time to rethink packaging, so that its merits, and not just its defects are kept firmly in mind. Packaging has had a poor press, but needs a profound reassessment. Innovation in the way in which prescription drugs are packaged could do a lot for society.

---

[5] Research2Guidance, a Berlin-based company that researches mobile IT, sees, too, a growing number of 'app stores' for mHealth. Applications will include those around fitness and nutrition. The firm also says that, in mHealth, multiple revenue streams will accelerate after 2012, based not so much on downloading applications as on mHealth services and hardware sales, buttressed by sales of drugs and other medical products, and—inevitably—by advertising [34].

References

1. Clean Up Britain (2011) Campaign launch press release. http://www.cleanupbritain.org/CLUB_launch_release.doc. This and all Internet documents in this paper accessed 23 June 2011
2. Hope C (2011) Packaging around toys to be slashed under government plans. The daily telegraph. http://www.telegraph.co.uk/earth/businessandecology/recycling/8569352/Packaging-around-toys-to-be-slashed-under-Government-plans.html. 10 June 2011
3. Waste Resources Action Programme (WRAP) (2010) Courtauld commitment 2. http://www.wrap.org.uk/retail_supply_chain/voluntary_agreements/courtauld_commitment/index.html Phase 2 targets. http://www.wrap.org.uk/retail_supply_chain/voluntary_agreements/courtauld_commitment/what_are_the_targets.html. 4 March 2010
4. Popova M (2010) Rethinking packaging. Bigthink. http://bigthink.com/ideas/19986. 7 May 2010
5. Woudhuysen J, Kaplinsky J (2009) Energise! A future for energy innovation. Beautiful Books, London
6. Department for Environment, Food and Rural Affairs (2010) Survey of com-mercial and industrial waste arisings 2010—final results. http://archive.defra.gov.uk/evidence/statistics/environment/waste/documents/stats-release101216.pdf. 16 Dec 2010
7. RecyclingPortal.EU (2011) England: provisional waste management statistics published. http://www.recyclingportal.eu/artikel/26072.shtml. 14 Feb 2011
8. Office for National Statistics (ONS) (2010) Mid-2009 population estimates: England; estimated resident population by single year of age and sex. http://www.statistics.gov.uk/downloads/theme_population/mid-09-uk-eng-wales-scot-northern-ireland-24-06-10.zip. 24 June 2010; ONS (2004) Mid-2004 population estimates: England; estimated resident population by single year of age and sex. http://www.statistics.gov.uk/downloads/theme_population/mid-2004-improved-migration-revised-13-05-10.zip. 13 May 2004
9. Horne R (2006) Compliance, adherence, and concordance: implications for asthma treatment. Chest 130:65S–72S. doi:10.1378/chest.130.1_suppl.65S
10. Lewis A (1997) Non compliance: a $100 billion problem. Remington Report 5: 14–15. This extra cost in the US is more than five times the size of the annual expenditure made on drugs by Britain's National Health Service
11. For 2006-7, the National Institute for Clinical Excellence (NICE) reported these costs as between £36 million and £196 million. NICE (January 2009) Costing statement: Medicines adherence: involving patients in decisions about prescribed medicines and supporting adherence. http://www.nice.org.uk/nicemedia/pdf/CG76CostStatement.pdf. Cited in Lithgow E (Spring 2010) Adherent Packaging. Pharmaceutical Manufacturing and Packing Sourcer. http://www.meadwestvaco.com/mwv/groups/content/documents/document/mwv023390.pdf
12. Elliott R (2009) Non-adherence to medicines: not solved but solvable. J Health Serv Res Policy 14:58–61. doi:10.1258/jhsrp.2008.008088
13. Ho PM et al (2006) Effect of medication nonadherence on hospitalization and mortality among patients with diabetes mellitus. Arch Intern Med 166:1836–1841 Cited in ibid
14. Ho PM et al (2006) Impact of medication therapy discontinuation on mortality after myocardial infarction. Arch Intern Med 166:1842–1847 Cited in Elliott, op cit
15. GBI Research (2009) The future of the pharmaceutical packaging industry—emerging economies poised to capitalize on increasing demand. Summary http://www.the-infoshop.com/report/gbi106802-pharma-packagi.html. Dec 2009
16. Woudhuysen J (2011) Big Pharma, small ambition. Spiked-online. http://www.spiked-online.com/index.php/site/article/10204/. 21 Feb 2011
17. Rivers PH et al (1998) Measurement of anticonvulsant adherence behaviour in the community using a medication events monitoring system (MEMS). Health Care Anal 6:308–316
18. Diaz et al (2001) Use of the medication event monitoring system to estimate compliance in patients with schizophrenia. J Psychiatry Neurosci 26(4):325–329. http://www.ncbi.nlm.nih.gov/pmc/articles/PMC167186/

19. Stora Enso (no date) Stora Enso in brief. http://www.storaenso.com/about-us/stora-enso-in-brief/Pages/stora-enso-in-brief.aspx
20. Stora Enso (2007) Stora Enso Pharma DDSi—for ultimate compliance. http://www.storaenso.com/media-centre/product-news/2007/11/Pages/stora-enso-pharma-ddsi-for-ultimate-compliance.aspx; Stora Enso (15 February 2011) Stora Enso Pharma DDSi Wireless—a new medical package offering improved opportunities for real-time adherence control. http://www.storaenso.com/media-centre/product-news/2011/02/Pages/stora-enso-pharma-ddsi-wireless-en.aspx; Penfold C (September/October 2010) How can intelligent packaging best aid patient compliance. Pharma. http://www.designcognition.com/wp-content/uploads/2010/10/10-09-29-Pharma-mag-Intelligent-packaging-Patient-Compliance-Chris-Penfold.pdf. 7 Nov 2007
21. Qolpac. Personal communication. See www.qolpac.com
22. MWV (no date) Cerepak electronic compliance packaging. http://www.meadwestvaco.com/HealthcarePackagingSolutions/SolidOralDoseandAdherence/MWV021872
23. Bang and Olufsen made Helping Hand before transferring the rights for the product to MWV. See slideshow (no date). http://medicomnordic.synkron.com/graphics/Videos/Flash%20files/medicom_slideshow_Helping_Hand.swf
24. Vodafone (no date) Mobile Patient Reported Outcomes (mPRO), and accompanying demonstration video. http://enterprise.vodafone.com/products_solutions/health_solutions/mpro.jsp
25. PricewaterhouseCoopers (2010) Biotech reinvented. Where do you go from here? http://www.pwc.com/gx/en/pharma-life-sciences/pdf/biotech-reinvented.pdf (registration re-quired)
26. Penfold C (2011) Medical device and packaging convergence. design cognition. http://www.designcognition.com/tag/biotech/. 11 March 2011
27. Jones E et al (2010) RFID pharmaceutical tracking: from manufacturer through in vivo drug delivery. J Med Devices 4(1):015001 1–015001 8. doi:10.1115/1.4000495
28. Searches conducted for the phrases 'electronic pack' and 'mobile health'. EUR-Lex. http://eur-lex.europa.eu/RECH_mot.do
29. Searches conducted for the phrases 'electronic pack' and 'mobile health'. EudraLex and Public Health. http://ec.europa.eu/health/documents/eudralex/index_en.htm and http://ec.europa.eu/health/ehealth/policy/index_en.htm
30. EU (2010) Regulation (EU) No 1235/2010 of the European parliament and of the Council. Official J Eur Union. http://ec.europa.eu/health/files/eudralex/vol-1/reg_2010_1235/reg_2010_1235_en.pdf. 31 Dec 2010
31. Europa press release (2011) Digital agenda: Kroes and Dalli welcome Council Presidency eHealth declaration on delivering better health care. http://europa.eu/rapid/pressReleasesAction.do?reference=MEMO/11/287&format=HTML&aged=0&language=en. 11 May 2011
32. Deloitte Center for Health Solutions (May 2011) 2011 Survey of health care consumers, global report: key findings, strategic implications. http://www.deloitte.com/assets/Dcom UnitedStates/Local%20Assets/Documents/ US_CHS_2011ConsumerSurveyGlobal_062111.pdf
33. Quittner AL et al. (2008) Evidence-based assessment of adherence to medical treatments in pediatric psychology. J Pediatr Psychol 33(9):916–919. doi:10.1093/jpepsy/jsm064. http://www.ncbi.nlm.nih.gov/pmc/articles/PMC2639495/
34. Research2Guidance (2010) Mobile Health Market Report 2010-2015: The Impact of Smartphone Applications on the Mobile Health Industry. http://www.research2guidance.com/shop/index.php/downloadable/download/sample/sample_id/62/. 11 Nov 2010
35. Deloitte Center for Health Solutions (May 2010) 2010 Survey of Health Care Consumers: Key Findings, Strategic Implications. http://www.deloitte.com/assets/Dcom-UnitedStates/Local%20Assets/Documents/US_CHS_2010SurveyofHealthCareConsumers_050310.pdf. Cited in Deloitte (14 September 2010) Issue Brief: The Mobile Personal Health Record: Technology-enabled self-care. http://www.deloitte.com/us/2010mobilepersonalhealthrecord

Chapter 2
The Four Horsemen: The Evolution of Science and Society

Michael Vloeberghs

Abstract This paper may have an unusual flavour. Although I show some of the medical advances of our time, I also present these from the perspective of a doctor who works as a surgeon and has observed society from different angles over the past 25 years. This paper is not meant to be patronizing or overly cynical, but I would like to think there are valuable observations and lessons just by looking at "evolution" of mankind. I chose this title, as in my opinion the Four Horsemen, Famine, Plague, War and Death, are the main factors that drive human evolution. I'll be reviewing the role of each individually by doing some time travel, setting the scene for the appearance of the horsemen, seeing how society-medicine has coped, or not, and will try to come to some outlook to the future.

Keywords Global industrialisation · Advanced medicine · Infection · Vaccines · History and society

2.1 Remember the Origins

A spark of intelligence led a rag tag band of hominids to get organized first in hunter gatherer groups, which already called for a fair amount of teamwork. This evolved further to agricultural and later urban communities.

This led to a schism in lifestyle between urban and agricultural communities. Trade appeared which set the social dynamics up to this day. The only major change is industrialization of every aspect of life, past, current and likely future.

M. Vloeberghs (✉)
Paediatric Neurosurgery, Nottingham University Hospital, Nottingham, UK
e-mail: michael.vloeberghs@nottingham.ac.uk

P. Breedon (ed.), *Smart Design*,
DOI: 10.1007/978-1-4471-2975-2_2, © Springer-Verlag London 2012

2.2 Famine

Eighty per cent of the world's population is urban as for some reason people assume life in cities is better and you can earn more money to buy food; rather than grow it yourself.

This last thought is naïve as food production is very variable. The appearance of food is seasonal, which was well known to my parents and grandparents. We are now upset if we do not have strawberries at Christmas.

In the past the hunger months were June-July; the supplies of the previous year were running out and the August harvest had not started yet. To make matters worse crops can fail, there are biblical pests and weather is unpredictable.

Madagascar is a powerful anecdote. Nearly all the farming land is taken up by crops meant to produce biofuel for behemoth companies. If the trend is continued they will need to import food. According to economists, this should not be a problem as the employment biofuel crops generate allows the locals to pay for the import.

Currently food production is globally industrialized. This does not mean the locals are better off or have easier access to food. Where in the past farming communities were relatively safe from starvation while cities were vulnerable, the opposite is currently true.

Realize the biggest famine is self-inflicted: dieting and eating disorders.

The biggest triumph of the food industry is diet food. For some reason, food with fewer calories is more expensive. The real solution is simpler: Just eat less of whatever you are eating, you will lose weight and as a bonus save money.

2.3 Plague

Sickness and disease have always been a part of life whether of bacterial, viral, prionic and other origin. There are effectively no "new infections"; we just have not discovered all of them yet. The cornerstone of infection control is general health, which in turn leads to good immunity and repair mechanisms. The discovery of infectious agents like bacteria and viruses is relatively new; finding an applicable cure even more recent.

Medicine evolved from an empiric discipline e.g. "Galenic" medicine, based on dogmatic knowledge and doctrine, to an observational discipline in the early eighteenth century. Jenner was one of the first to use a directed preventative measure.

He noticed that milkmaids and cow handlers did not get smallpox. They did get cowpox, a more or less innocuous infection by a closely related virus. Having no idea of what he was doing, Jenner scraped the blisters from the milkmaids and scratched the skin of non-infected people; protecting them from smallpox. Clearly, Jenner did not have to do an ethics application to do this experiment and produced the "scarification" technique and "vaccination." Vaccination contains the Latin for cow: "vacca."

The same century Anthony Van Leeuwenhoek, by fluke develops the microscope, which in the nineteenth century allows the birth of microbiology with Pasteur and Koch (Tuberculosis).

Infection control has always been an issue and is also the main reason surgery did not evolve as rapidly as Medicine. Operations were possible in a limited way, but often the patients died of secondary infections.

Up until recently, women were at risk of contracting "childbirth fever," which is an infection of the womb after delivery. Before antibiotics, this condition was usually fatal.

In the mid nineteenth century a reputed obstetrician noticed his patients contracted "childbirth fever" when his assistants did the delivery after coming from cadaver dissection. He advised hand washing after touching the cadavers and the infection rate dropped by 70%.

Antibiotics became available in the 1920s, which led to a huge improvement in general health of the population and extended longevity. This also led to a dramatic political conclusion in the UK. Bevan assumed, in his original manifest for the creation of the NHS in 1948, that most of the life threatening diseases, which in the day were infectious, would be controlled and no thought was given to the Four Horsemen, who would find other opportunities.

Extended longevity gives new opportunities to Plague and Famine. We are now confronted with prion disease, ebola, H1N1, obesity, heart disease, diabetes and dementia.

2.4 War

Injury is a part of life too and so is the struggle to survive. Our bodies are only to a certain extent self-healing and only cope with very limited damage thresholds. Our ancestors quickly noticed that penetrating injuries were far more efficient in taking out the competition. If you penetrate the skin, you can die of acute blood loss and if that did not work, the secondary infection of the wound would do it. This led to a perpetual race between offensive devices and protection. Sticks and stones to spears and arrows to crossbows, until Marco Polo brought efficient gunpowder back from China along with the first banknotes and pasta.

2.5 Society of Rapid Change

We take a step away from the narrative and look at society. Radical changes happened over time. The industrial evolution of the eighteenth and nineteenth centuries set a new matrix for society. Another technological boost occurred in the 1930s and WW2, leading directly to the second technology wave of the 1960s and the tertiary wave of the late twentieth century, with ever increasing speed.

The industrial revolution began with automated looms, which precipitated poverty in urban areas and created a new dependence and an opportunity for Famine: the need for employment to buy food. The ability to feed yourself or your family depended entirely on your ability to produce.

Urban proximity gave new opportunities for Plague. People migrated to cities where there was no provision of clean water or sewerage. Abject social conditions and poor health led to rampant tuberculosis and social disease like alcoholism.

The industrial revolution also gave War an interesting opportunity. War was now an industry, but the military leaders were not prepared for this. Outdated tactics of troop deployment were used in the industrialized artillery battles of WW1 leading to mass destruction.

During WW1, advances in medicine were made to cope with the overwhelming amount of injured. A fairly new condition appeared; gas attacks, which damaged the soldier's lungs, leading to further development of ventilators. Even then there was a shortage of medical equipment and soldiers sang: "They're gassing us and there's one ventilator for the four of us!"

Application for the X-rays discovered by the Curies came of age leading to trucks called "the little Curies" carrying mobile X-rays equipment on the battlefield.

2.6 WW2 Technology Boost

The destruction of WW1, which had seriously gotten out of hand, led to the punishing post war conditions put to Germany. The conditions also implied scrapping research into weapons, no military research or organizing of any kind. No one said anything about, alternative fuel, rocket engines, combustion engines or materials to the benefit of mankind. This "sleeper science" along with the right political climate and a people with a feel for organizing made astounding progress. Where initially the Versailles treaty was adhered to, it was quickly forgotten when Hitler came to power, and stayed there. Near unlimited funding was given to engineering research and resurrection of industry with a replay of WW1 in mind.

Ferdinand Porsche along with Hitler designed a car, wagen in German, destined to be affordable for the German people, volk in German. Put both together and you get: Volkswagen. The Volkswagen Beetle came to exist. Ferdinad Porsche also designed the Tiger Tank, which ravaged Europe.

In the late 60s and early 70s, fuel injection in motorcars was seen as an innovation, despite direct fuel injection being used in the Messerschmidt 109 since the late 30s.

Werner Von Braun and the SS created the V1 and V2 weapons. The V2 was a challenge. A vertical lift device that would have a sub-atmospheric flight, where there was no oxygen and to get there you needed a volatile and light fuel.

The solution was to use ethanol produced from grain combined with turbine pumps that mixed the ethanol with liquid oxygen. The down side was the combustion temperature was so high the metal steering flaps melted on the prototypes. The solution was the creation of carbon flaps.

In 2003, 60 years after the V2, Virgin Galactic did a much-publicized sub-atmospheric flight, using the same principles. We are now told biofuel is an innovation and that carbon is an effective replacement for metal.

A less known race between the US and the Nazis was that for nuclear power, Tritium vs. Uranium, which culminated in the detonation of two nuclear bombs to avoid the need for invasion of the Japanese mainland and save Allied lives.

2.7 The Second Wave

The economic boost represented by the reconstruction of Europe and the appearing of new technologies led to a second wave. The post war climate created the space program and the need for observation and listening devices. The golden age of the 60s had arrived. The space race, telecommunications via Telstar, but also the tension of the cold war, in which the balance between East and West was partially restored by the Rosenbergs. Microelectronics became implantable and Medtronic produces the first Cardiac Pacemaker.

Communications, particularly after the appearance of the transistor, played an ever-increasing role. Television, embryonic in the 1940s in Germany, gave a near non-stop feed to people. The Vietnam War was televised leading directly to the protests of the late 60s.

With delay and because of government control, communications were slower to evolve in the East but ultimately led to the collapse of the Soviet Union, starting in Gdansk in 1980.

2.8 Tertiary Wave

The appearance of home computing gave everyone the computing power previously reserved to academics and the military. Arpanet turned into Internet and allows anyone to do or see anything at near light speed.

A current trend is a return to biology. Once the number crunching was done, attention turned to individual genes, mastering biological code and interfering with stem cells.

If measuring progress by direct applications to improve quality of life or tackle common disease is done, you will find progress is slow.

Although creating publicity there are few direct applications, which are usually limited to rare diseases.

The astute person is likely to have asked him- or herself what the point of this historical dialogue is.

There are cycles in time and a common denominator. The industrial revolution was driven by, well industry, and the profit to be made of influx of often colonial funds.

In the 1930s, Nazi politics were funded by industry and the Nazis ultimately took over the state structure and funding.

The US space program was heavily funded to send a political message to the East.

This day and age governments are bankrupt and can no longer afford spectacular shows of engineering, medicine or war. Governments find they need to look at private funding again to fund the previously mentioned.

The contrast in social dynamics and lifestyle, result of the industrial revolution, has never been replicated.

An example is the space program; at a massive expense, an inflated V2 was built by the same team. The Apollo program made it to the moon but the missions were narrow minded without lateral thinking. I can only imagine what difference the discovery of water on the moon would have made then.

To make matters worse, we have not come any further. The space shuttle was a step back to solid fuel boosters, leading to the Challenger disaster and currently the world needs to rely on outdated, disposable, 60s Russian rockets to maintain the ISS.

The bottom line is; the money dried up and the private sector is again a convenient partner but fickle partner. Private industry only funds, preferably low risk, rapid return projects. This also means the rate of progress and what will progress is determined by the deliverables in profit.

2.9 Medical Industry

I have been a surgeon for 25 years and have dealt with medical industry in several fields. Currently my clinical practice is the surgical treatment of movement disorders in children. I do this with either implantable drug delivery pumps or Deep Brain Stimulation (DBS).

These systems originated in the 1980s and actually have remained unchanged since. The only real change in either system is the battery life, which has tripled over 20 years; evolution driven by the mobile phone industry, not by medical requirement.

The current drug delivery pumps have a programmed death to force the user to replace the pump and are not used to their full potential. There are a variety of drugs that could be used for a variety of conditions but this is of little interest to the corporations.

Mary Shelley came up with the idea of Deep Brain Stimulation mid nineteenth century by writing "Frankenstein's monster." Although spectacular and a great treatment in adults and children, the primary indications for DBS are rare and so medical industry is artificially expanding the uses of the same device.

Medical industry follows the cycle I previously described. They build on old established technology, make minor changes that do not need regulatory review, re-license the same device for new applications to recover investment and maximize profit.

Medical industry is overseen by several organizations. Often a first step is CE marking, which allows a product to be sold and used in Europe, which is a lesser risk that doing this in the US.

The gold standard is the US Food & Drug Administration approval but this can take 10–15 years and involves clinical trials that need to be funded by the company seeking approval.

The World Health Organization also has regulatory capacity and has recently come under scrutiny for corruption during the "flu pandemic."

In the UK there is an extra filter on top of all the others: the National Institute for Clinical Excellence (NICE). NICE is an organization that responds very slowly to requests and essentially disguised cost cutting. NICE was taken to court by the breast cancer patients because NICE refused to approve Herceptin, the drug that really makes a difference.

I would like to take you through two different approaches to medicine. One is the management of head injury and the other is space medicine.

After a head injury, usually the result of a road traffic accident, we pile medical technology onto the patient.

The patient will be ventilated, given artificial coma, have a pressure monitoring device, receive Mannitol etc. Despite all these measures there is no change in the outcome of head injury. Head injury remains a devastating problem with dire consequences. There is some research funding for the management of head injury and medical industry looks favourably at this common disorder because it involves equipment and devices.

Space medicine looks at injury in a completely different way. The space program has its roots in the military with an emphasis on risk avoidance as in industry, who use the sometimes absurd HAZOP protocols. This approach is necessary as there are limited options for emergency care in space and evacuation is hazardous. An injured person is unlikely to survive re-entry.

The lesson to be learned is prevention. Avoid having a head injury by wearing a seatbelt, looking after your vehicle, do not drink and most of all do not be stupid.

A back to reality message is that "high end" medical technology is of no use to most people, you will never need it. The biggest contributions to health in the past have been sanitation and access to clean water.

What most people will be confronted with is heart disease, obesity, high blood pressure, in essence lifestyle diseases, with directly impact on longevity. As in head injury, you are for a large part responsible for contracting these lifestyle diseases.

What does longevity mean? Humans have a limited display date and this increased dramatically in the twentieth century. The mean was 35 years around 1900 and is currently around 80 years in Western Europe, –7 years in the US because of the absence of reliable healthcare. The maximum appears to be 120 years after which the system is worn out and the repair mechanisms are exceeded. If you want to live up to 120 you will need to invest in you.

In Zen Buddhism you may be asked to solve a Koan, which is a reputably unsolvable question: what is the sound of one hand clapping or does a falling tree make noise when there is nobody there?

The real answer to the Koan is: the question does not make sense, which is a very useful motto for every researcher. The answer to the Koan of longevity is not, do you want to live longer but, would you like a better life.

Eighty per cent of the world could do with an improvement of their quality of life, whatever this means to the individual.

2.10 Conclusion

Where does this leave us as individuals? Hitler said it was a great advantage to leaders that men did not think. This motto made it possible for a criminal organization to overtake one of the most advanced and powerful countries in the world, with all mayhem ensuing. Our society is heavily industrialized, industry feeds us, entertains us, dictates which products we use socially or medically and we face corporatism, without appeal.

Human nature has not changed. We have a very short span of attention and want quick fixes; an attitude which led to the collapse of the banking system. One overwhelming novelty is that the planet has acquired a nervous system. The Internet has given us awareness.

Far from trying to pass a "New Age" message, I believe that in view of the current state of things we should individually and generally ask ourselves the right questions, educate, collaborate, plan for longer term, think laterally, question corporatism and use common sense. Do you need to drive, do you need to eat all this, drink all this, should we build unprotected nuclear reactors in earthquake zones?

We need to realize that our actions and thoughts are now interconnected at each level. Each level duplicates the other; whether we fight off a virus infection or take to the streets in protest it is a form of response. The awareness allows people to compare lifestyle, thoughts and social dynamics and make up their own mind. Just think of the role of Facebook in the Egyptian revolution and Wikileaks.

Interdisciplinary programs are an essential in this setting. The psychologist has input for the engineer's project as will the doctor and the designer. A mathematical solution may solve a medical problem. There are ideas hidden on computers and solutions on scraps of paper in drawers that will never see the light of day if we do not take initiative and leave it all to industry or grant giving bodies.

2.11 Death

Finally; there is one horseman missing: Death. Death is what you call a "Dude." Death can be relaxed about everything, does not need to stress about finding new opportunities like War, Plague or Famine because every time, everything, everywhere, dies.

Bibliography

1. Barnes J (2009) A history of the world in 101/2 chapters. Vintage, NY
2. Bryson B (2004) A short history of nearly everything. Black Swan, London
3. Chandler D (2011) Research update: improving batteries' energy storage
4. New method allows a dramatic boost in capacity for a given weight (2011) MIT News. http://web.mit.edu/newsoffice/2011/better-battery-storage-0725.html. Accessed 10 Jan 2011
5. DARPA (Defense Advanced Research Projects Agency) (2012) About our work. http://www.darpa.mil/. Accessed 12 Jan 2012
6. Hobsbawm E (2009) On empire: America, war, and global supremacy. New Press, New York
7. Gombrich EH (2008) A little history of the world. Yale University Press, New Haven
8. Graves R (2011) Overview of the innovative clinical and medical sectors advances, trends and issues. Horizon Scanning lecture series, Nottingham Trent University. 21 Sept 2011
9. The Kings Fund (2012) Public health. http://www.kingsfund.org.uk/topics/public_health/index.html. Accessed 10 Jan 2012
10. Michener J, Lattimer RL (1985) All we did was fly to the moon. Whispering Eagle Press, Gainesville
11. Mitchell WJ, Inouye AS, Blumenthal MS (eds.) (2003) Beyond productivity. National Academies Press, Washington
12. Quennell M Quennell C (series) A history of everyday things in England. Batsford, London
13. Shelley M (2003) In: Hindel M (ed) Frankenstein. Penguin Classics, London
14. Sternfeld A (1959) Soviet writings on earth satellites and space travel. MacGibbon and Kee, London
15. The Wellcome Trust (2012) Medical history and humanities. http://www.wellcome.ac.uk/. Accessed 12 Jan 12

Chapter 3
Changing the Fabric of Society: An Exploration of Smart Textiles in Product Design

Jonathon Allen

Abstract There has been a flurry of activity in the area of smart textiles over the past decade or so [Baurley (Per Ubiquit Comput 8:274–281, 2004); Eves et al. (New nomads: an exploration of wearable electronics. 010 Publishers, 2010); Garcia (ed) (Architextiles. Wiley-Academic, 2006); Lupton (Skin: surface, substance, design. Princeton Architectural Press, 2002)], and we are now beginning to see the transition from protypical items, to commercial applications. Yet the uptake of textiles in product design is relatively poor. Whilst there are some notable exceptions [Baurley (Per Ubiquit Comput 8:274–281, 2004); ElekTex (http://www.elektex.com, 2009); Philips (http://www.lumalive.philips.com, 2011)] the majority of applications remain prototypical. Yet why is this, and what can smart textiles offer the field of product design? This paper interweaves some of the developments and applications of smart textiles, drawing examples from other domains, such as art, architecture, engineering and bio-medical fields, as well as materials science, in the process, exploring future applications of smart textiles in industrial and product design.

Keywords Smart textiles · Product design · New materials

J. Allen (✉)
School of Humanities and Communication Arts, University of Western, Sydney, Australia
e-mail: j.allen@uws.edu.au

P. Breedon (ed.), *Smart Design*,
DOI: 10.1007/978-1-4471-2975-2_3,

3.1 Introduction

Textiles gave birth to the industrial revolution, taking cottage production of cloth to mechanical production on a massive scale. In the process the textile industry fast-tracked other industries, setting a trajectory to the future, and in turn changed the very fabric of society. Textiles are not just the stuff of clothes, apparel, and home furnishings though. Today textiles have radically changed other industries, from electronics to construction, medicine to architecture. With the development of smart materials—materials that can either *sense*, *react* or *adapt* (or any combination of these) to their environmental conditions or to other stimuli [1]—a new field of smart textiles promises to cause further radical changes across a broad range of industries, and its impact on society "has the potential to transform many aspects of our lives" [2].

3.2 Smart Textiles and Smart Use of Textiles

There has been a flurry of activity in the area of smart textiles over the past decade. Philips Design in 2000 explored the idea in their project and book, *New Nomads: an exploration of wearable electronics by Philips* [3]; Ellen Lupton's *Skin* [4] in 2002 showcased some fascinating work; and in the UK between 2004 and 2007 the *Smart Textile Network* [5] was established as a think tank linking research in academia and industry. Meanwhile, a number of universities, particularly in Europe and the US, and companies such as Philips have their own research programs devoted to smart textiles [6, 7].

There are already examples of incorporating electronics into clothing on the market. For instance, UK-based company, ElekTex® [8] produce an electro-conductive textile that is lightweight, soft and flexible, yet highly durable. The textile is also washable, and can work over a temperature range of −40 to +70°C so provides a great choice for apparel and clothing. From a men's suit by leading UK retailer Marks and Spencer, to cycling shorts by Pearl Izumi, to snowboard jackets by O'Neill, a number of fashion houses have incorporated the ElekTex® touchpad—a fabric-based controller designed for use with an iPod—in their clothing ranges.

The fascination with textiles has not just been confined to clothing and apparel—textiles have been embraced by artists, architects, and engineers, providing inspiration to challenge and explore the realm of the possible. Artists such as Anish Kapoor, in collaboration with the engineering firm Ove Arup and Partners, stretched the boundaries of fabric in his installation sculpture, *Marsyas*—an extraordinary 150 m long red trumpet-like installation consisting of three metal rings connected by a PVC coated polyester skin—that filled the Turbine hall in London's Tate Modern gallery [9]. Our understanding of three-dimensional geometry is enriched by works such as this. This point has not been missed by architects, many of whom, with the aid of advanced computational software and textiles, have explored the unique forms and surfaces that textiles can deliver. In 2006 a book, *Architextiles* [10], was devoted to this theme. In more humble ways, textiles have also been readily adopted by the construction industry—be it geotextiles used in civil engineering projects to prevent erosion and stabilise ground works, or breathable fabrics such as Tyvec® [11] used in many building projects. Much of the transport industry relies on textiles, be it for engineering or interior applications. In aircraft and marine craft, much of the structure of the vehicle is based upon woven composites, and as light-weighting and fuel-efficiency become increasingly important in the automotive sector, more structural components and panelling are adopting woven composites over metal predecessors. More obviously, textiles have always had an important role in the interior of vehicles, and there is a great deal of research and testing of new textiles for interior applications, be it fabrics that wick away sweat from the driver, or are stain-resistant, or can heat or cool to provide maximum comfort for the driver, textiles have an acknowledged and important place in these industries.

More recently, BMW's Geometry and functions In 'N' Adaptations (GINA) concept car has explored shape-shifting fabric as the vehicle exterior [12]—an inspirational design and perhaps the promise of things to come—yet whilst this might be a smart application of textiles, it in itself is not utilising smart textiles. But, smart textiles could automatically perform the shape-shifting of the vehicle exterior in response to speed—using a woven or knitted matrix of pressure-sensing piezo-electric fibres that generate electricity in proportion to air pressure (*ergo*, speed of the vehicle), shape memory alloy filaments that can change the form of the fabric, in addition to fibre-optic filaments that can transmit data through the fabric, along with woven fibres of carbon, carbon nanotubes and aramid for strength—a smart vehicle skin could not only be lightweight, but have a form of material intelligence that responds automatically to its conditions [13].

3.3 Smart Textiles in Products: Expanding Senses and Sensibilities

But what of product design? Has this field embraced the potential of textiles? There are some stand out areas—mainly in the realms of sporting goods, furniture, and accessories for electronic components—but there are few commercial examples of products. The ElekTex® soft keyboard, designed by Sam Hecht [14] in 2001 is a notable exception, and later IDEO's collaboration ElekTex® and Logitech in 2004 saw the *Logitech Keycase* keyboard win an Industrial Design Excellence Award; but by and large the foray into consumer electronics has not really been taken up since. This is understandable given the research and development focus on interactive displays for mobile and computer devices over the last few years, but smart textiles may yet come of age.

There is now great interest in the ability to screen-print a number of smart films directly onto fabric, or to directly weave conductive fibres with electroactive polymers (materials that can move or give off coloured light for instance), so that dynamic screens can be seamlessly integrated into fabrics. A number of fabrics have incorporated either electroluminescent wires, fibreoptics or discreet LEDs to light up the material in the dark or else to produce a flexible display. For instance, Philips have produced their *Lumalive* [6] fabric and their research division are looking at the application of OLEDs in textiles and products, and Kodak have a research program dedicated to flexible displays and textiles [15].

Whilst most consumer electronics consist of simple input devices (a keyboard, button or switch for instance), some electronic signal-processing, and simple output devices (such as a screen, or an audio device), a smart textile can be both an input device and an output device. Further, the power required to operate a device could actually be generated by the fabric itself. Professor Zhong Lin Wang and his colleagues from Georgia Tech have developed a fabric that incorporates zinc oxide nanowires coated in tetraethoxysilane (a weatherproofing protective coating) that can generate electricity from small motions [16]. Whilst the fabric does not produce a lot of electricity—about 80 milliwatts per square metre—this can be sufficient to power portable electronic devices such as MP3 players or medical implants. This gives new meaning to power dressing and power walking—going for a walk in your shirt and suit could actually power the mobile office.

A smart fabric can also be used to regulate temperature, airflow, or moisture levels, by tightening or loosening threads in the weave and weft of the fabric to open up a gap in the material. For clothing there are obvious advantages—providing more or less insulation to maintain heat or to cool the person, or venting the clothing to help reduce sweat build up (the fabric could also have an active deodorising system, whereby the material can release micro-encapsulated perfumes at set conditions, triggered by heat or an electric current for instance). Regulating temperature of an electronic product could also benefit from the use of smart textiles. Typically heat sinks, consisting of a highly conductive material with a large surface area, are used to cool computer chips. As the processing power of computer chips increases and they get a lot smaller, they also generate more heat and so, paradoxically, the larger the heat sink needs to be to cool the chip. Smart fabrics offer the potential to address this problem in three ways. Firstly the material can be designed to change and increase its surface area or increase airflow through the material as required, in the same way described above. Secondly, either thermoelectric cooling textiles that incorporate Peltier elements that transfer heat from the inside to the outside of the fabric, or Phase Change Materials that absorb and later dissipate heat energy by changing form can be used. Finally the layout of the electronics upon the fabric itself can consist of a more distributed processing unit (instead of a Central Processing Unit (CPU) a Distributed Processing Unit can perform the same functions), much like how a neural network operates.

This final opportunity is incredibly exciting for the benefits it could offer in how we interact with technology. Textiles can be embroidered or printed upon using conductive materials, and so flexible circuitry can now relatively easily be incorporated onto a fabric. The difficulty up until recently has been to ensure the circuitry can flex without damage, but advances in processing technologies and coatings has now addressed this. As electronic components are getting smaller and smaller, they too can be incorporated into a fabric, and so the time of truly wearable electronic devices is now here.

Whilst there are plastics that can be impregnated with scents (e.g Auracell® by Rotuba [17]), fabrics have for millennia had the capacity to carry and give off smells (sometimes not desirably!). So smart textiles can also have this ability—whether pre-impregnated with a particular smell, or washed or perfumed from time-to-time. Further, in the same way that conditioners and coatings can be applied to fabrics by washing them, so too with most smart textiles. This offers an interesting opportunity for the ultimate customisation of worn technologies—these products can be coloured, perfumed, softened and textured, or treated to be hypo-allergenic, simply by laundering. Given that our sense of smell can be an important recal of memory, and is also a useful way of triggering an alert response (the smell of smoke or gas typically draws an immediate response whilst awake), the safety applications are also of value. A worn device that could provide warning signals by tightening (using shape-memory polymers or alloys, or a hydrogel); vibrating (piezo fibres) or changing temperature (a phase change material or a simple resistant wire), or even give off a distinct smell (triggered by heat for instance) could expand the way we currently interrelate with technology.

The sensory opportunities that smart textiles afford design, especially the tactile and haptic qualities, are of real interest—particularly as a way of interacting with technology. Some smart textiles can change temperature, colour, or their pliancy, so the range of tactile feedback can be more greatly attuned than current product interfaces. Sharon Baurley, Head of Design at Brunel University, and former coordinator of the *Smart Textiles for Intelligent Consumer Products* [5, 18] program, has explored the opportunities to augment remote interpersonal communication and emotion with tactile feedback. Her *Emotional Wardrobe* [19] project has produced a number of prototypical garments, such as *Communication-Wear* that augments a mobile phone with tactile feedback using electronic textiles that can touch/stroke the wearer (gentle squeezing) by incorporating shape memory alloys into the fabric. In essence, a reassuring phone call or a message to a loved one could come with a hug or a pat on the back.

In a similar fashion, Philips Research has developed *The Emotions Jacket* [7] for enhancing the cinema experience, so that the wearer can literally feel some of what the characters on screen do. The applications from their research into the sensory experience will no doubt begin to manifest in home entertainment and gaming devices in the not too distant future. Such an immersive world is no longer the stuff of science fiction but thanks, in part, to smart textiles is progressively becoming real. Smart textiles offer us the opportunity to expand our sensory interaction with technology, and could also change the way we think of products, and in so doing blur the boundary between technology and us.

3.4 Conclusion

There are some significant problems of utilising smart materials for product application, however, as identified by Oliver and Toomey, "The key challenge is turning the designs into real products and systems, a result of the restricted availability of active materials in design friendly formats. ... A current challenge is obtaining sufficient quantities of materials ... to be able to incorporate into implementable design led products" [2].

Whereas fashion, architecture, art and engineering domains can afford to create the bespoke, and indeed only one-off items may be required, product design has a culture of designing for mass production. This is perhaps what we are seeing in regard to the application of smart textiles—the circumstances have not yet been right for product design, but incredible advances are being made. This is perhaps the pivotal turning point for smart textiles. The Industrial Revolution was predicated on the ability to mass-produce textiles; in turn, industrial design as a discipline was formed, and mass production became the *modus operandi* of product design and manufacture. Mass-producing smart textiles will see a new revolution, and perhaps textiles once again will change the fabric of society.

References

1. Tao X (2001) Smart technology for textiles and clothing—introduction and overview. In: Tao X (ed) Smart fibers, fabrics and clothing, 1–6. Woodhead Publishing. http://www.knovel.com/web/portal/browse/display?_EXT_KNOVEL_DISPLAY_bookid=819&VerticalID=0. Accessed 16 May 2011
2. Oliver R, Toomey A (2010) Design drives materials innovation. Ingenia 42:38–43
3. Eves D, Green J, van Heerden C, Mama J and Marzanao S (2001) New nomads: an exploration of wearable electronics. 010 Publishers, Rotterdam
4. Lupton E (2002) Skin: surface, substance, design. Princeton Architectural Press, New York
5. Smart Textiles Network. http://smarttextiles.co.uk. Accessed 14 April 2011
6. Philips (2011) Lumalive. http://www.lumalive.philips.com. Accessed 14 April 2011
7. Philips Research (2009) Philips emotions jacket—a new level in immersive cinematic experience. http://www.research.philips.com/technologies/emotionsjacket/index.html. Accessed 14 April 2011
8. ElekTex (2009) ElekTex smart fabric. http://www.elektex.com. Accessed 14 April 2011
9. Kapoor A (2002) Marsyas. Installation in the tate gallery, London 9 Oct 2002–6 April 2003. http://www.tate.org.uk/modern/exhibitions/kapoor/default.htm. Accessed 14 April 2011
10. Garcia M (ed) (2006) Architextiles. Wiley-Academic, London
11. DuPont (2011) DuPont™ Tyvek®. http://www2.dupont.com/Tyvek/en_US/index.html. Accessed 14 April 2011
12. BMW TV (2008) The BMW GINA light visionary model. http://bmw.tv/intl/us/video.do?articleID=6368&spaceID=4&tagID=33. Accessed 14 April 2011
13. Allen J (2008) Materials: the incredible lightness of being. Curve 24:44–46
14. Museum of Modern Art (2010) The collection: Sam Hecht. The museum of modern art, New York. http://www.moma.org/collection/artist.php?artist_id=22582. Accessed 14 April 2011

15. Stephenson S, Johnson D, Kilburn J, Mi X-D, Rankin C and Capurso R (2004) Development of a flexible electronic display using photographic technology. Eastman Kodak company, Research and Development Laboratories, Rochester. http://www.kodak.com/US/plugins/acrobat/en/corp/flexDisplay/flexdisplayTechPaper.pdf. Accessed 14 April 2011
16. Qin Y, Wang X, Waang Z (2008) Microfibre-nanowire hybrid structure for energy scavenging. Nature 451:809–813. doi:10.1038/nature06601
17. sAuracell (2011) Auracell: a scented polymer by Rotuba. http://www.auracell.com/. Accessed 14 April 2011
18. Baurley S (2004) Interactive and experiential design in smart textile products and applications. Per Ubiquit Comput 8:274–281. doi:10.1007/s00779-004-0288-5
19. Baurley S, Stead L (2007) The emotional wardrobe. In: Inns T (ed) Designing for the 21st century: interdisciplinary questions and insights. Gower, Farnham

Chapter 4
Smart Housing: An Intelligent Environment for Aging Independently

Marilyn Bruin, Sauman Chu, Jodene Riha, Amanda Smoot and G. Mauricio Mejia

Abstract The purpose of *The Smart House, Livable Community, Your Future* project was to demonstrate how good design, assistive technology, and well-planned community amenities, support independent living as well as increase usability across the lifespan. We used a variety of strategies to share information on smart design to Baby Boomers. A 1,300 square foot exhibition, designed as a home, allowed visitors to tryout assistive technology and observe structural innovations. Workshops explored innovative design for (a) consumers and caregivers; (b) developers, remodelers, contractors, and designers; and, (c) community planners and policy makers. A *Smart House, Livable Community, Your Future* website provides free, downloadable activities and information to Baby Boomers and encourage community conversations. The paper describes various aspects of the project and summarizes participants' reactions to the content; the paper also explores implications for designers and planners interested in creating intelligent environments. The project inspired participants to think about strategies for improving the built environment. Multiple outreach formats, an interactive exhibition, workshops, and an interactive website, helped us to reach the Baby Boomer cohort.

Keywords Independent living · Interactive websites · Universal design

---

M. Bruin (✉) · J. Riha · A. Smoot
Housing Studies, University of Minnesota, Minnesota, USA
e-mail: mbruin@umn.edu

S. Chu · J. Riha
Graphic Design, University of Minnesota, Minnesota, USA

G. M. Mejia
University of Minnesota and University of Caldas, Columbia, USA

P. Breedon (ed.), *Smart Design*,
DOI: 10.1007/978-1-4471-2975-2_4,

4.1 Introduction

As Baby Boomers, the 78 million Americans born between 1946 and 1964 who comprise nearly 26% of the U.S. population, become empty-nesters and imagine post-work lifestyles they will change the demand for housing, public transportation, recreation, and community-based programs [1, 2]. Overall, Boomers are healthier, wealthier, better educated, expect to live and work longer, and play harder than their predecessors [2]. The majority (75%) believe they will be able to stay in their current homes as they age [3]. However, despite a predilection to deny the effects of aging, it is likely Boomers will eventually experience many of the physical and cognitive changes, including sensory and motor impairments, experienced by their elders [4]. If so, they will likely need to modify their current home or move to accessible housing; only 5% can expect to age in their long term home without modification [4]. However, with declining home values and shrinking retirement savings, many Boomers may lack the financial resources to self-finance home modifications, in-home supportive care and services, or moves to senior housing [5, 6]. Furthermore, Baby Boomers' tendency to not plan for their futures, and to assume that the market will meet their demands, has challenging implications for housing, services, and other aspects of community design.

4.2 Intelligent Strategies in the Built Environment

Adapting the home environment may be a cost-effective solution for Baby Boomers who want to remain independent and engaged with their social networks [7]. In the United States, home modifications are defined as changes to the living environment that increase ease of use, safety, security, and independence. Common modifications that increase comfort, safety, and accessibility, for individuals dealing with the physical and cognitive issues that often accompany aging, include redesigning the first floor with a bedroom and bath, climate controls within optimal reach, non-slip floor coverings, bathrooms with grab bars, and accessible showers, sinks, and toilets. A no-step entrance allows access for visitors with mobility issues. Emergency alert systems help individuals call for assistance in the event of a fall or other emergencies.

Accessibility solutions do not need to be overly complicated nor expensive; even small adaptations can make a significant differences in the ability of the resident to navigate their home environment safely and independently [8]. There is a significant need for consumer education, assistance, and funding options to encourage the wider incorporation and acceptance of home modifications, universal design, and visit ability features that enable all people, regardless of ability, to live better and more independently [7–9].

Boomers searching for new housing, typically view the community and the neighbourhood as important as the housing itself [10]. "Simply put, an 'age-friendly' community promotes the physical and psychological well-being of community members throughout the life cycle" [11, p. 8]. In other words, developing age-friendly, vital, and livable communities that offer affordable, appropriate, and vibrant housing options with accessible transportation, access to health care, social interaction, cultural and religious opportunities attracts and retains Boomers as well as everyone who follows them [3, 5, 11]. Planners and developers need to encourage the development of accessible housing in all areas of the community to meet the future needs of aging Boomers [10]. New housing should be designed and current housing retrofitted to incorporate elements of universal design, visit ability, and accessibility that allow residents to navigate their homes safely and independently, as well as engage in a community that encourages participation in social and civic life [3, 7, 12]. Once planners and developers understand how to develop and maintain age-friendly housing and communities, they also need to market the housing characteristics, services, and community amenities that exemplify age-friendly supportive housing and communities.

4.3 Developing Content

We referred to frameworks, set goals, and wrote a back story about a leading-edge Baby Boomer couple to help guide the selection of objects and the design of the exhibition as well as the development of workshops and website content to exemplify a home that was aesthetic, comfortable, safe, and fully accessible across the lifespan. The primary framework was Quality of Life Basics, design features and objects that combine to enhance a sense of worth and well-being for residents [13]. We strove for an environment that emphasized privacy, belonging, control or ability to use the environment, and a sense of safety and security for residents and visitors. In addition to the physical structure, we considered psychological concepts such as home as a setting for nurturing relationships with intimate others, a place to develop and maintain social networks, self-identity, and personalized space, to maintain continuity, as well as to serve as a base of activity and family/childhood memories [14]. We wanted to demonstrate home modifications that enhance the home and protect its equity value, reduce stress, add comfort, preserve physical energy, provide options to meet changing needs, and facilitate ongoing interests and social activities. Finally, we selected examples that were widely available.

4.4 Innovative Outreach

The purpose of *The Smart House, Livable Community, Your Future* project was to demonstrate how good design, assistive technology, and well-planned community amenities, can support independent living by increasing usability, safety, and quality of life across the lifespan. The information was delivered through a variety of formats and intended to educate visitors about design principles, assistive technology, and integrative inclusive community planning.

4.4.1 Exhibition

A 1,300 square foot exhibition, open February 5–May 23, 2011 in the Goldstein Museum of Design, was designed as a small, attractive home (see Figs. 4.1, 4.2, 4.3) inhabited by fictional, leading-edge Baby Boomers. The exhibition represented the outcomes of the renovation of a one-level, suburban ranch built in the 1960s to support an active, physically- and socially-engaged lifestyle in the twenty-first century. Visitors were encouraged to try out the objects and furnishings in the exhibition, starting with a sturdy, comfortable bench in a neighbourhood park setting. A welcoming no-step entrance exemplified how accessibility and good design combine to enhance the front of a house. Visitors could sit on comfortable furniture, try a "lift-assisted" chair, and experience good lighting, as well as examples of high contrasts in colour and texture that ameliorated the effects of changes in vision. In the home office, visitors sat in a fully adjustable desk chair at an ergonomically designed desk. The universally designed kitchen with accessible appliances and cabinets facilitated meal preparation from a seated position; visitors were encouraged to handle easy-to-use-kitchen utensils, open drawers, and pull inserts out of the cabinets. The bath featured a walk-in shower, comfortable, accessible toilet and sink as well as an example of a reinforced wall for grab bars. Visitors learned about an alert system, auto-dispensers for medications, special environmental controls, and tools and technologies that allowed the Boomer couple to do the activities they enjoyed, and stay connected to the world. For example, a home computer with SKYPE™ helped them stay in touch with family overseas; software on the computer helped the husband track changes in blood pressure and weight, and identified concerns and shared information with his physician. Brief videos and notes helped visitors understand the motivations of the couple to remodel their home. Visitors observed structural innovations for bathrooms, doorways, kitchens, wall colours, and lighting that increased safety and comfort, as well as demonstrated good design and aesthetics.

4.4.2 Workshops

Workshops presented by the co-authors explored innovative design for homes and age-integrated communities for three audiences: (a) consumer-focused workshops for Boomers, seniors, and caregivers; (b) developers, remodelers, contractors, and designers; and (c) community planners and policy makers. Community members and professional experts provided critical guidance in planning the design

Fig. 4.1 Welcoming entrance and park setting

Fig. 4.2 Kitchen area in the exhibition

workshops for specific audiences and related the exhibition to their needs. Workshop participants were encouraged to access the website and use resources to generate local interest and on-going conversations about home modification and independent living, as well as to advocate for community planning that will increase accessibility and safety for aging Baby Boomers. Exhibition visitors and

Fig. 4.3 Bedroom design in the exhibition

workshop participants were encouraged to complete a qualitative, open-ended on-line survey; comments from the survey provided evaluation data describing the experiences of exhibition and workshop participants.

4.4.3 Website

A study conducted by Chadwick-Dias et al. [15] compared web interaction between seniors and adults. Seniors had more difficulties performing interactive activities; so, websites for elders should contain clear navigation and links, provide redundant cues, and present instructions on how to perform specific tasks. Additionally, Hilt and Lipschultz [16] concluded that older adults rely on search engines to navigate the web, and the perception of credibility (given by the interface) affects the interaction. Recently, in a comprehensive study about computer usage by older adults, age was correlated to negative attitudes and anxiety toward the use of computers [17].

Seniors used information resources that were valuable for them, and they used a limited range of functionalities. The primary usage of computers by seniors was for social communication, health and education information access, and entertainment. Furthermore, seniors who received initial training on how to use the computer had higher self-confidence and satisfaction when they were asked to use the computer again [15]. Baby Boomers are the first generation with wide experience using digital interactive technologies in the workplace [16]. Indeed, Harris and Edelman [17] pointed out that Baby Boomers are passionate about computers and the Internet. Compared to previous generations Baby Boomers have different behaviours and attitudes toward using the Internet because they have experience with computers and highly innovative technologies in the workplace [18]. The above studies, as well as our usability tests, provided a foundation as we created and designed the website to showcase and accompany the exhibition. Our usability findings suggested similar outcomes as Chadwick-Dias et al. [15]. Providing a clear description and instructions on what each particular link's connections was definitely preferred by the participants.

The *Smart House, Livable Community, Your Future* website [http://smarthouse.design.umn.edu/] presents content from the exhibition and program series as well as interactive learning modules for aging-related home planning. The learning modules were created by students in a game design course, instructed by one of the co-authors. The goal of the learning modules is to provide educational and engaging environments for users to learn how to improve different settings at their homes. In addition, the website provides free downloadable activities and information to facilitate community conversations about universal design, home modification, and community design. The local public television station produced a 30 minute broadcast as well as seven brief videos that were incorporated into the exhibition and website. The 30 min broadcast and videos are available on the website.

4.5 Outcomes

Innovative outreach strategies were tested to convey messages to our target groups. Through short videos, Jim and Sarah, the imaginary residents, introduced and set a context for each vignette in the exhibition. Through written notes, exhibition visitors were privy to Jim and Sarah's challenging decision-making process, as well as the motivations and emotions involved in transforming one's home. Museum visitors were encouraged to actively engage, experience, and enjoy the space; they were instructed to sit down, play with objects, and try tools. Visitors seemed to appreciate the interactive exhibition and workshop activities. Survey comments suggested the participants appreciated the interactive nature of the exhibition and the workshops.

> The way it was laid out as it would be in a real home. It is better to see it like it would appear rather than in pictures.
> Visual! Great examples/exhibits and hands on experience.
> The hands on approach and the ideas generated by seeing it. Good use of technology!

We hoped to inspire visitors to adopt technology and design solutions, including assistive technology and home modifications, as well as community design that promote independent living and community integration across the lifespan. The following comments indicated that the exhibition, workshops, and website met these goals.

I liked the way it makes the features of aging in place practical - so that those who may say 'I don't need that' can see the benefits.
Showing how you can downsize and still do it attractively. Also, the importance of green space, parks etc. nearby.
I liked the entry way ideas because that is the area of the home I never considered being a problem for an older person.

Participants learned new information and thought about its applications. Some mentioned they will "file it away in their memory for possible future use," share it with others including contractors, and try to integrate it into current living and work environments. Individuals mentioned plans to broadly share the information, and one individual mentioned incorporating it into occupational therapy training. Participants seemed to appreciate that information was readily and publically available.

Actually, it was seeing that many of the things that make life easier are ready to purchase.
I liked the fact that it was put on by the University of Minnesota.
Maybe even try to set up some sort of program at my church to talk about home accessibility issues.
It's outreach to people who maybe don't know or have thought little about it.

Several respondents were not impressed; however, their comments affirm our focus on easy, practical, smart solutions often by using readily available objects.

I guess I was expecting something more complicated, some ideas that people might not think up on their own.
I didn't know what to expect, but didn't see anything very surprising. So without having clear expectations I was nonetheless a bit disappointed in the lack of "wow" factor.

4.6 Conclusion

Creating or retrofitting housing to meet the needs of all individuals, regardless of ability, is an option to allow older adults to remain in their homes and maintain autonomy and privacy. Although many modifications and adaptations can be easily installed for a relatively low price, it is clear that many older adults are confused regarding effectiveness and cost [8, 17]. Such hesitation can include an inability to identify how or where to incorporate modifications, how to hire professionals, how to pay for adaptations, and the uncertainity about how changes will affect the value or the property [12].

The survey comments confirmed that participants want to maintain Quality of Life Basics, specifically control or ability to use the environment, and a sense of safety and security, in their homes. Furthermore, they seem to value the psychological concepts of home as a setting to maintain continuity, as well as a base of activity. Although difficult to interpret, perhaps the comment noting the lack of a "wow" factor in the exhibition reflected individual's assessment that practical and simple design strategies did not fit her sense of self.

Overall, we feel the project inspired participants to rethink home and community, and to adopt intelligent strategies in designing their home environment. The use of multiple formats, an interactive exhibition, workshops, and an interactive website, helped us to reach the Baby Boomer cohort who seem comfortable receiving information in multiple and innovative formats. The visual displays in the exhibition seemed to help visitors inculcate the concepts.

References

1. Greenblatt A (2007) Aging baby boomers. CQ Researcher http://library.cqpress.com/cqresearcher/document.php?id=cqresrre2007101900&type=hitlist&num=0. Accessed 3 Aug 2011
2. HUD (2005) Elderly housing consumption: Historical patterns and projected trends. Housing and Urban Development, Washington
3. AARP Public Policy Institute (2010) Strategies to meet the housing needs of older adults. http://assets.aarp.org/rgcenter/ppi/liv-com/i38-strategies.pdf. Accessed 3 Aug 2011
4. Sabia JJ (2008) There's no place like home: a hazard model analysis of aging in place among older homeowners in the PSID. Res on Aging 30:3–35
5. Myers D, Ryu SH (2008) Aging baby boomers and the generational housing bubble: foresight and mitigation of an epic transition. J Am Plan Assoc 74(1):17–33
6. Pynoos J, Liebig P (2009) Changing work, retirement, and housing patterns. Generations 33(3):20–26
7. Pynoos J, Nishita CM (2003) The cost and financing of home modifications in the United States. J Disabil Policy Stud 14(2):68–73
8. Zola IK (1997) Living at home: the convergence of aging and disability. In: Lanspery S, Hyde J (eds) Staying put: adapting the places instead of the people. Baywood Publishing Company, Amityville
9. Lanspery S, Callahan JJ, Miller JR, Hyde J (1997) Introduction: staying put. In: Lanspery S, Hyde J (eds) Staying put: adapting the places instead of the people. Baywood Publishing Company, Amityville
10. Wylde MA (2008) Right house, right place, right time: community and lifestyle preferences of the 45+ housing market. BuilderBooks, Washington
11. Scharlach AE (2009) Creating age-friendly communities: why America's cities and towns must become better places to grow old. Generations 33(2):5–11
12. Memken J, Earley N (2007) Accessible housing availability for the growing U.S. elderly population. Hous Soc 34(1):101–115
13. Wylde M, Baron-Robbins A, Clark S (1994) Building for a lifetime: The design and construction of fully accessible home. The Taunton Press, New Town Connecticut
14. Hayward DG (1978) An overview of psychological concepts of home. In: Brauer RG (ed) Priorities for environmental design research. EDRA, Washington
15. Chadwick-Dias A, McNulty M, Tullis T (2003) Web usability and age: how design changes can improve performance. Conference on universal usability, ACM, Vancouver, British Columbia, pp 30–37
16. Hilt M, Lipschultz J (2005) Mass media, an aging population, and the baby boomers. Erlbaum, Mahwah
17. Harris L, Edelman M (2006) After sixty: marketing to baby boomers reaching their big transition years. Paramount Market Publishing, Ithaca
18. Wagner N, Hassanein K, Head M (2010) Computer use by older adults: a multi-disciplinary review. Comput Hum Behav 26(5):870–882. doi:10.1016/j.chb.2010.03.029

Chapter 5
Enhancing Visibility in Bicycling Apparel Using Integrated Flashing LEDs

Helen Koo and Lucy Dunne

Abstract The purpose of this study is to inform the development of visibility-enhanced bicycling safety clothing using flashing LEDs on bio-motion joints, and to evaluate the effects of this active visibility technology on bicyclist visibility under both daytime and night-time conditions. To evaluate the effectiveness of LED visibility aids, a survey and video evaluation of visibility was conducted online with 133 participants, who both drive a car and ride a bicycle. Video recordings of bicycle trials were made under four viewing conditions: with the flashing LEDs on during daytime and night-time, and with the flashing LEDs off during daytime and night-time. Data were analyzed with ANOVA and Duncan Test, and the results show that the bicycling smart wear improved the perceived visibility of the bicyclist at night, and helped the bicyclist to be more recognizable and detectable under night-time conditions ($p < 0.001$). With respect to observer perception regarding safety and accidents, wearing the developed bicycling smart wear during times of poor visibility increased perception that bicycling is safer ($p < 0.001$). The results revealed the LED technology improved perceived visibility of the bicyclist, observer's recognition and detection of the bicyclist.

Keywords Keywords Bicycling · Smart clothing · LEDs · Visibility

H. Koo (✉) · L. Dunne
The Wearable Technology Lab College of Design,
University of Minnesota, 240 McNeal Hall 1985 Buford Ave Saint Paul,
San Francisco, MN 55108, USA
e-mail: kooxx044@umn.edu

P. Breedon (ed.), *Smart Design*,
DOI: 10.1007/978-1-4471-2975-2_5,

5.1 Introduction and Backgrounds

When bicyclists ride outside, they are exposed to dangerous situations, especially during darkness, due to decreased visibility (distance or time at which an object can be clearly discerned) [1–3]. Low visibility threatens bicyclists, particularly as the speed of automobiles and motorcycles is much faster than that of bicycles. By the time drivers see bicyclists, it is often too late to stop [2, 3]. Increasing the bicyclists' visibility considered as important to protect bicyclists from accidents. Therefore, many researchers and community guidelines suggest that bicyclists increase their visibility by wearing clothing that aids visibility during dark and foggy conditions [1–6]. Garments such as vests or jackets, and accessories like helmets or belts aimed at increasing visibility generally use passive visibility aids (retro-reflective or brightly coloured materials that reflect light) or active visibility aids (luminescent materials that generate light). Passive visibility is well studied, but active visibility is a newer and less-studied concept. Here, we are concerned with evaluating the benefit of active visibility using LEDs. Most bicycling clothing includes LEDs attached on the middle torso, usually on the back of the body. However, the human ability to perceive an object as a human (rather than a street sign or another object) is strongly related to the movement of joints. Many researchers have found that biological motion configurations enhance observers' recognition of bicyclists and pedestrians [7–12]. The purpose of this study is to inform the development of visibility enhanced bicycling safety clothing using flashing LEDs on the joints of the body, and to evaluate the effects of this clothing design on bicyclist visibility by comparing the ability of vehicle drivers to visually detect moving bicyclists under both day- and night-time conditions.

5.2 Research Methods

To evaluate the visibility-enhanced bicycling safety clothing, a female bicyclist participant was videotaped wearing a black jumper and pants, black shoes, and a black helmet. This clothing was augmented with magnetic flashing LEDs. To eliminate any potential colour factor that might affect the bicyclist's visibility, the subject wore all black, with no retro-reflective trim. The yellow colour of lamps, flashing lights, and retro-reflective materials has been shown to enhance detection and recognition by drivers at night likely because the human eye is most sensitive to yellow [2]. Therefore, in this study, yellow and red magnetic flashing LEDs (thatscoolwire.com; MGFL-RB12) with a diameter of 10 mm and a height of 13 mm and blinking in 0.2 s intervals were utilized. To evaluate visibility, the cyclist described above was filmed by a passing car driver. (In order to limit possible differences in viewing conditions among the bicycle trials, the four trials described here were filmed, and the films were then used for the subsequent usability tests, in which the test subjects viewed the films, rather than the actual bicyclists.)

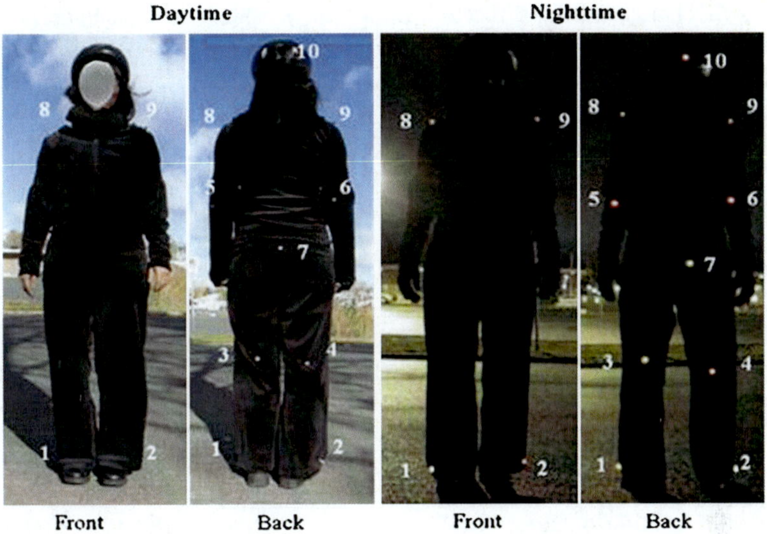

Fig. 5.1 Configuration of flashing LEDs

Fig. 5.2 The four video clips (VCs: VC1, VC2, VC3, and VC4 from *left* to *right*)

A digital camera (Canon IXUS 870 IS) was tightly fastened to a helmet worn by the driver with rubber bands. The camera was fixed in a vertical direction, from the forehead to the eyes, in order to record a view similar to that of the driver. The flashing LEDs were attached to the black suit (the jumper and the pants) as well as the helmet. Ten points of LED attachment were selected in consideration of enhancing the visibility of the various body segments and joint movements: 1-left ankle, 2-right ankle, 3-left knee, 4-right knee, 5-left elbow, 6-right elbow, 7-hip, 8-left shoulder, 9-right shoulder, and 10-head (see Fig. 5.1).

The video recordings of bicycle trials were made fewer than four viewing conditions (VC): during daytime without clothing LEDs (VC1); during daytime with clothing LEDs (VC2); during night-time without clothing LEDs (VC3); and during night-time with clothing LEDs (VC4) (see Fig. 5.2). The site, time of day (which will change between the two conditions), speed, and distance were the same in each recording situation in order to eliminate any possible interference from these four elements. In all four bicycling trials, the bicycling distance was 1800 feet (approximately 550 m) on the right shoulder of a relatively flat, two-lane roadway; the speed of the vehicle following was 20–21 mph; the bicycle speed was 20–22 mph; and the video recording was collected when the vehicle was situated 45–50 feet (approximately 14–15 m) behind the bicyclist.

After the video clips were recorded, a survey and visibility evaluation was conducted using Amazon Mechanical Turk, a resource through which users are paid to complete human-intelligence tasks (https://www.mturk.com). The purpose of the survey was to solicit observers' impressions about the bicyclist's visibility during the day and at night. 133 subjects(age; $m = 29.07$, $SD = 8.85$, range 18–58), 76 men and 57 women, who drive a car and also ride a bicycle, participated in the survey and evaluated visibility by viewing the video recordings and giving their impressions regarding the visibility of the bicyclist under day- and night-time conditions. A total 15 questions were asked including demographic questions, evaluation of perceived visibility in the video clips, and relative evaluation of video clips. Open-ended, multiple choice and 7-point Likert scale (1 = "strongly disagree", 7 = "strongly agree") questions were used. Each video clip had duration of approximately 1 min, and the total usability test session lasted about 10–15 min. To avoid Turk users "gaming" the system from influencing the results [13], two "test" questions with quantifiable correct answers were included to determine whether or not the observer was actually watching clips: these asked the number of the bicyclists in the video clips and whether LEDs existed or not. Responses that answered these questions wrong were rejected.

5.3 Data Analysis and Primary Results

Observers were asked to use 7-level Likert scale to indicate level of perceived visibility (the visibility of the bicyclist was very visible), recognition (I was able to immediately recognize that there was a bicyclist), detection (I was able to readily detect flashing lights on the clothing of the bicyclist), flashing LEDs (the flashing lights on the clothing helped improve the visibility of the bicyclist) and configurations (the locations of the different flashing lights on the clothing helped me recognize that it was a bicyclist that I saw in the video clip) of the bicyclist in VC1, VC2, VC3, and VC4. Differences were analyzed with ANOVA and Duncan Test.

The results revealed significant differences in the perceived bicyclist's visibility, observer's recognition and detection of the bicyclist, improvement of the visibility by using flashing LEDs and configurations of the flashing LEDs. The data analysis results are presented in Table 5.1 and Fig. 5.3.

With respect to observer perception regarding safety, observers were asked to use the 7-level Likert scale, 1 for strongly disagree and 7 for strongly agree, to indicate whether or not they would feel safer if they were to wear clothing with flashing LEDs.

Table 5.1 The ANOVA analysis result for the observers' impressions on the four VCs

Analysis	VC_{Group}	Mean	SD	F-value
Visibility	$VC1_2$	5.41	1.25	330.11*
	$VC2_1$	6.23	1.24	
	$VC3_4$	1.92	0.77	
	$VC4_3$	4.47	1.40	
Recognition	$VC1_2$	6.17	1.22	195.75*
	$VC2_1$	6.51	1.13	
	$VC3_4$	2.76	1.72	
	$VC4_3$	5.14	1.45	
Detection	$VC1_3$	1.48	1.38	517.78*
	$VC2_2$	1.49	1.47	
	$VC3_4$	1.42	1.21	
	$VC4_1$	6.55	1.06	
Flashing LEDs	$VC1_3$	1.38	1.17	330.11*
	$VC2_3$	1.34	1.18	
	$VC3_2$	1.71	1.64	
	$VC4_1$	6.39	1.25	
Configurations	$VC1_2$	1.53	1.38	316.01*
	$VC2_2$	1.47	1.33	
	$VC3_2$	1.67	1.52	
	$VC4_1$	2.65	2.37	

(*$p \leq 0.001$)

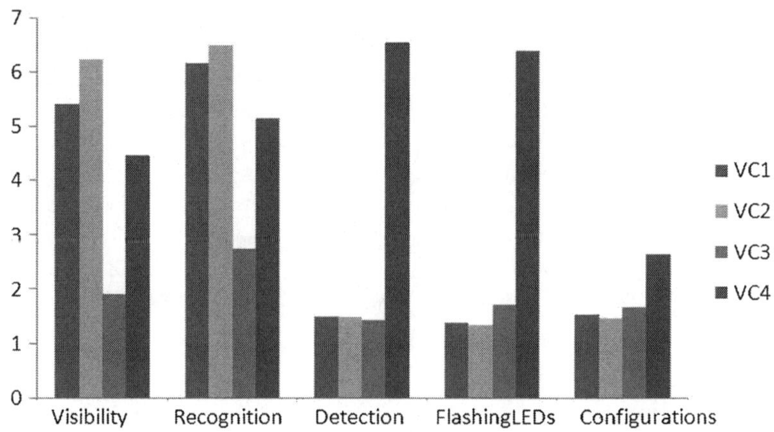

Fig. 5.3 The ANOVA analysis result for the observers' impressions on the four VCs

The result indicated that wearing the flashing LEDs on clothes during times of poor visibility ($m = 6.53$, $SD = 1.13$), increased impressions that bicycling is safer, compared with conditions of good visibility ($m = 2.81$, $SD = 1.94$).

The observers were also asked use the same scale, 1 for strongly disagree and 7 for strongly agree, to indicate whether or not they feel or anticipate that they are less likely to be in an accident if they wear clothes with the flashing LEDs. The survey result showed that wearing the flashing LEDs on the clothes during times of poor visibility ($m = 6.48$, $SD = 1.34$), as opposed to during conditions of good visibility ($m = 3.17$, $SD = 2.12$) increased impressions that bicycling has a lower risk of accidents. All of the observers agreed that increasing the visibility of bicyclists would help to reduce bicycle accidents especially during dusk or night-time conditions to the yes or no questions.

5.4 Discussion

In analyzing the observer's *perception* of the bicyclist's visibility, the flashing LEDs seemed to improve the visibility of the bicyclist more at night-time than in daytime. The flashing LEDs increased the visibility of the bicyclist in both day- and night-time conditions. However, the differences of the bicyclist visibility between VC3 and VC4 are bigger than the VC1 and VC2. That would indicate that using flashing LEDs at night-time increased the bicyclist's visibility more than using LEDs during the daytime (an expected result, due to the light of the LEDs being relatively weaker than sunlight and therefore mainly visible when they are in shadow). A similar result was recorded for the observers' *recognition* of the bicyclist. At night it was hard for viewers to recognize the bicyclist immediately. The bicyclist in VC4 was more immediately recognizable than the bicyclist in VC3, and similarly, the bicyclist in VC2 is more immediately recognizable than the bicyclist in VC1. However, again the difference between VC3 and VC4 was more than VC1 and VC2, demonstrating a similarly more significant impact of the LEDs in daylight than at night.

There were no flashing LEDs on the bicyclist in VC1 and VC3, and the observers were not made aware of this fact. Still, the observers answered correctly by responding that the flashing LEDs were not detectable in VC1 and VC3. The results indicated very little differences among the observers' *detection* of the bicyclists in VC1, 2, and 3. In contrast, the flashing LEDs in VC4 under night-time conditions were approximately four times more readily detectable.

5.5 Conclusion

In the study described here, wearable flashing LEDs increased the bicyclist's visibility and safety, with a corresponding decrease in perceived detection time. The most effective usable conditions for flashing LEDs for bicyclists are during the night or in poorly lit conditions. While this evaluation shows that LEDs do improve visibility over no protection, we have not yet evaluated the utility of LEDs in comparison to more established means of providing visibility (particularly in dark conditions), such as retro-reflective materials. In future studies, it would be beneficial to perform similar tests using more observers and with more variation in design parameters like location and type of LED lighting, and compare other visibility aid materials.

References

1. European Road Safety Observatory (2008) Pedestrians and cyclists. ERSO. http://www.erso.edu. Accessed 16 Jan 2011
2. Kwan L, Mapstone J (2004) Visibility aids for pedestrians and cyclists: a systematic review of randomised controlled trials. Accid Anal Prev 36:305–312
3. Rumar K (1990) The basic driver error: late detection. Ergonomics 32(10–11):1281–1290
4. Bloomberg D, Hale A, Preusser F (1986) Experimental evaluation of alternative conspicuity-enhancement techniques for pedestrians and bicyclists. J Saf Res 17(1):1–12
5. McGuire L, Smith N (2000) Cycling safety: injury prevention in Oxford cyclists. Inj Prev 6:285–287
6. National Highway Traffic Safety Administration (2007) Traffic safety facts: 2007 data: bicyclists and other cyclists. National Center for Statistics and Analysis. http://www-nrd.nhtsa.dot.gov/pubs/810986.pdf. Accessed 17 Jan 2011
7. Balk SA, Carpenter TL, Brooks JO et al. (2006) Pedestrian conspicuity at night: how much biological motion is enough? In: 2006 annual meeting of the transportation research board. http://www.thecochranelibrary.com/userfiles/ccoch/file/Safety_on_the_road/CD003438.pdf. Accessed 17 Jan 2011
8. Luoma J, Penttinen M (1998) Effects of experience with retroreflectors on recognition of night-time pedestrians: comparison of driver performance in Finland and Michigan. Transp Res 1(1):47–58
9. Luoma J, Schumann J, Traube E (1996) Effects of retroreflector positioning on nighttime recognition of pedestrians. Accid Anal Prev 28(3):377–383
10. Owens A, Antonoff J (1994) Biological motion and pedestrian safety in night traffic. Hum Factors: J Hum Factors Ergon Soc 36(4):718–732
11. Tyrrell A, Wood M, Chaparro A, Carberry P (2009) Seeing pedestrians at night: visual clutter does not mask biological motion. Accid Anal Prev 41:506–512
12. Wood M (2005) Limitations in drivers' ability to recognize pedestrians at night. Hum Factors: J Hum Factors Ergon Soc 47(3):644–653
13. Kittur A, Chi H, Suh B (2008) Crowdsourcing user studies with mechanical turk. In: Proceedings CHI'08, Florence, Italy, pp 453–456. doi.10.1145/1357054.1357127, 5–10 April 2008

Chapter 6
Hydrophobic Smart Material for Water Transport and Collection

Robert H. Morris, Shaun Atherton, Neil J. Shirtcliffe, Glen McHale, Tilak Dias and Michael I. Newton

Abstract *Stenocara Gracilipes* (the Namib Desert beetle) is a desert dwelling beetle which has adapted to make use of fog as an alternative water source in an environment which receives little rain water. Using a combination of hydrophobic and hydrophilic areas on its carapace, the beetle is able to collect condensation on its back which is then channelled towards the mouth. In this paper we attempt to mimic this effect by selectively altering the hydrophobicity of a number of water repellent fabrics. Fabrics were treated using Granger's Extreme Wash-in to make them hydrophobic and then laser etched to alter the hydrophobicity. We show a clear relationship between the hydrophobicity of the fabric and the laser energy applied to the surface. Laser etching was used to create a herring bone pattern of channels on the surface of the fabrics. Water sprayed onto the surface preferentially followed the channels into a collection vessel, giving a collection efficiency of 81%. To replicate real world conditions dry ice was used to create fog which was then blown, using an electric fan, onto the fabric at a speed of approximately 2.5 km/h. The water vapour condensed on the surface and then followed the channels into a collection vessel. It was found that the patterned fabrics achieved a collection rate of 0.31 l h^{-1} m^{-2}.

Keywords Hydrophobicity · Biomimetics · Fog collection

R. H. Morris (✉) · S. Atherton · N. J. Shirtcliffe
G. McHale · T. Dias · M. I. Newton
Nottingham Trent University, Nottingham, UK
e-mail: rob.morris@ntu.ac.uk

P. Breedon (ed.), *Smart Design*,
DOI: 10.1007/978-1-4471-2975-2_6,

6.1 Introduction

Water repellent or hydrophobic surfaces are present in many natural systems from plants to animals. Humans have been studying and replicating these surfaces for a variety of applications for hundreds of years to add water repellency to man made systems. In this paper, we present a material which has been designed to use a variety of water repellent areas to facilitate the movement of water over its surface. This allows both for the preferential direction of water on surfaces and also for the collection of fog in arid environments.

6.2 Hydrophobicity

The propensity of a surface to repel water is known as hydrophobicity. The origins of this effect lie both in surface chemistry and topology. A water droplet on a hydrophobic surface will tend towards a sphere. As hydrophobicity increases, it is possible to reach a state known as superhydrophobicity [1, 2] at which point water droplets begin to form actual spheres which can roll on the surface.

It is possible to produce a hydrophobic surface by using materials in which the surface molecules act to repel individual water molecules such as fluorinated polymers, for example PTFE. Surface texture can enhance this tendency. At the interface between a gas and a liquid, the surface tension in the liquid will cause the formation of a 'drop'. It is for this reason that on most flat surfaces a water droplet has a hemispherical shape. In the same way as a person can remain supported on a Fakir's Carpet (bed of nails), hairs or other structures on a surface can be used to support a water drop, as long as they are spaced sufficiently closely. If the space between these surface features is filled with gas (or is a vacuum), the surface tension in the droplet will again cause the water to form a curved edge. As this droplet is now surrounded by gas on all sides, it will tend towards a sphere. This once again results in a hydrophobic surface. By combining these two techniques (surface texture and chemistry) it is possible to make a superhydrophobic surface (see Fig. 6.1).

There are countless examples of systems such as these in nature. Amongst the most well known is the Lotus leaf. Despite floating in muddy pools of water, the leaves of plants in the genus *Nulembo* (Commonly known as lotus) remain perfectly clean and dry. The surface of the Lotus leaf is covered in a microscopically bumpy surface further covered in waxy crystals. This combination of a textured and waxy (water repellent molecules) surface results in a superhydrophobic layer. As well as keeping the leaves dry, this texture also allows the surface of the leaves to be self-cleaning. As a water droplet rolls across the surface of the leaf, it collects dust and debris which has collected on the surface. This technique has been employed in several commercial technologies [3, 4].

Hydrophobic systems are not however confined solely to plants, *Stenocara Gracilipes* (the Namib Desert Beetle) (Fig. 6.2) uses a combination of hydrophobic and hydrophilic (the opposite of hydrophobic) surfaces to collect water

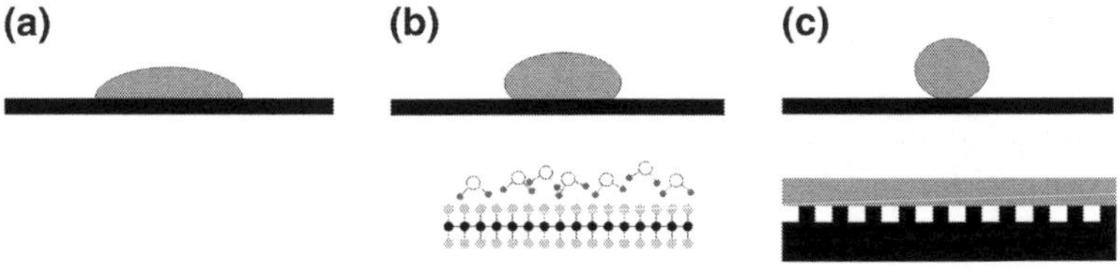

Fig. 6.1 Surface makeup determines the degree of observed hydrophobicity. A water drop in (**a**) partially wets the surface. The same drop in (**b**) now begins to take on a less spread shape due to interactions with a chemical layer. The same drop in (**c**) however tends towards a sphere as the surface is both chemically active and textured

Fig. 6.2 Namib desert beetles. Photograph by Thomas Schoch, reprinted under the CC-BY-SA-3.0 license

from early morning fog in desert areas. The abdomen of this beetle is covered in hydrophilic bumps surrounded by hydrophobic channels. The beetle stands facing an oncoming wind with its abdomen raised above its head. As the water molecules in the fog hit the hydrophilic bumps, they begin to accumulate. Once they reach a critical size, they detach from the surface and roll along hydrophobic channels to the mouth of the beetle. In this way, it is possible for the beetle to survive in environments with as little as 40 mm of rain per year [5]. By mimicking this system, in which areas of differing hydrophobicity are used to transport water, novel smart materials can be developed for a range of applications.

In order to assess the hydrophobicity of a surface, a measurement known as contact angle is often used. As the surface tension of a water drop causes it to tend towards a spherical shape when on a superhydrophobic surface and to a hemisphere on a hydrophobic surface, the angle between the surface plane and the bottom of the drop provides an indication of the relative hydrophobicity of that surface. This angle is known as the contact angle and is assessed using image processing of back lit drop photographs. In this project, a Krüss Drop Shape Analysis system is used to determine the contact angle of small water drops.

6.3 Technology

In this paper we present a novel smart material which uses regions of differing hydrophobicity to transport water. The material itself must be sufficiently flexible and lightweight that it can be used in remote environments for fog collection. It must also be sufficiently robust that it will withstand characteristic desert conditions. The weave of the fabric provides the necessary texture to produce a greater degree of hydrophobicity in combination with the chemical coating. This surface is then smoothed using a laser cutting machine (40 W M-300, Universal Laser Systems) to produce less hydrophobic regions within the greater hydrophobic background. The pattern of hydrophobic areas are applied to mimic that seen in plants which use a similar technique to transport water, as reported in [6]. The spacing of this pattern is optimised to ensure that maximum transport efficiency is achieved.

6.4 Surface Preparation

Sheets of fabric were purchased either pre-coated with polyurethane or were coated with a commercially available wash in waterproofing chemical preparation (Extreme Wash-in, Granger's, UK) by first soaking the material in an aqueous solution of the preperation during agitation for 90 min and then heating to 100°C for 1 h. The hydrophobicity of the surface was then assessed using the contact angle of water drops on its surface.

6.5 Material Selection

Initial testing focused on determining the optimum fabric for use with the system.

A variety of materials with a variety of weaves were purchased with and without hydrophobic coatings already applied to them. It was found that the greatest contact angle was with 100% acrylic whilst the lowest was with mixed natural fibres (Table 6.1). Fabrics purchased ready coated with polyurethane were found to have similar contact angles to those treated with Granger's Extreme Wash-in (Table 6.2) and therefore did not warrant the additional cost.

The laser system was used to remove the surface layer of the fabric only, resulting in a flatter profile and therefore a reduction in hydrophobicity. Since it is the greatest change in contact angle which is desired for this study (whilst maintaining hydrophocity), polycotton offered the best variation. Figure 6.3 shows contact angle against laser speed (Power = 3 W, pulses per inch = 1,000) for the chosen material (polycotton) a poor variation in contact angle (cotton) and a high variation resulting in a loss of hydrophobicity entirely (stormwear cotton).

Table 6.1 Contact angles of materials as purchased. Top table is purchased uncoated, bottom table is purchased coated with polyurethane

	Natural/synthetic	Contact angle (degrees)
Uncoated material		
3 oz cotton	Natural	Hydrophilic
3 oz polycotton	Synthetic	Hydrophilic
3 oz polyester	Synthetic	Hydrophilic
5 oz polyamide	Synthetic	127
15 oz tight weave acrylic	Synthetic	138
Polyurethane coated material		
7 oz nylon	Synthetic	134
12 oz nylon	Synthetic	138
15 oz cotton	Natural	140
3 oz polyester	Synthetic	140
5 oz polyester double coated	Synthetic	144
5 oz acrylic	Synthetic	146
Tight weave acrylic—thick	Synthetic	149
12 oz polyester	Synthetic	131
12 oz cotton	Natural	145

Table 6.2 Contact angles of previously uncoated materials after application of Granger's Extreme Wash-in

Material	Natural/synthetic	Contact angle (degrees)
3 oz cotton	Natural	137
3 oz polycotton	Synthetic	148
5 oz polyamide	Synthetic	136
3 oz polyester	Synthetic	133
15 oz tight weave acrylic	Synthetic	142

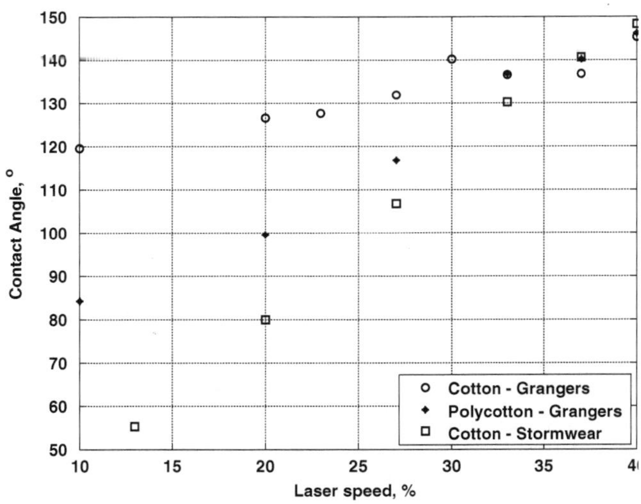

Fig. 6.3 Contact angle as a function of laser speed for three materials

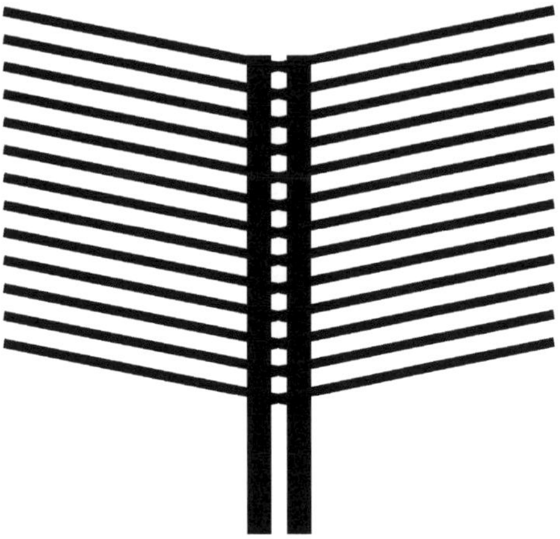

Fig. 6.4 Pattern of low hydropobicity patterned onto hydrophobic fabric

Table 6.3 Collection efficiency of patterned polycotton surfaces

Finger angle	Collection efficiency (%)
10	81
26	80
40	64

6.6 Surface Testing

The pattern shown in Fig. 6.4 was laser engraved onto the surface of the fabric in a variety of finger angles to determine the optimum pattern.

The surface of each spacing of the pattern was misted with water along with a control sample. The volume of water which reached the bottom of the central branch was measured. The volume of water collected is used with the volume of water applied to determine the efficiency of the surface as a water collection device. The results for the different pattern angles are shown in Table 6.3. It is apparent that the greatest efficiency is seen for an angle of 10°. This pattern was then used to determine the suitability for fog harvesting.

6.7 Fog Harvesting

The spacing determined in the previous experiment was subjected to a fog produced by immersing solid carbon dioxide chips in hot water. A fan was used to introduce a wind speed of approximately 2.5 km/h replicating conditions seen in the Namibian desert. The fabric was placed perpendicular to the oncoming wind and the water volume collected was measured over a period of an hour. For the first 30 min, no water was actually collected as it begins to form droplets on the fabric surface, after this time however it was found that water is collected at a rate per unit area of 0.35 l $h^{-1}m^{-2}$.

6.8 Conclusion

We have demonstrated that a pattern of hydrophobic channels on a hydrophobic sheet can be used as an easily transportable device for fog harvesting in arid environments which is capable of harvesting 10 ml of water per hour per square meter.

Acknowledgments We gratefully acknowledge the support of the EPSRC under Cross Disciplinary Feasibility Account EP/I016414/1.

References

1. Shirtcliffe NJ, Atherton S, Newton MI, McHale G (2010) An introduction to superhydrophobicity. Adv Colloid Interf Sci 161:124–138. doi:10.1016/j.cis.2009.11.001
2. Roach P, Shirtcliffe NJ, Newton MI (2008) Progess in superhydrophobic surface development. Soft Matter 4:224–240. doi:10.1039/b712575p
3. STO (2005) Technical Data Sheet 3206/INT.EN. http://www.stoshop.co.uk/assets/pdfs/TDS-StoLotusan-Color.pdf
4. Blossey R (2003) Self-cleaning surfaces—Virtual realities. Nature Mater 2:301–306. doi:10.1038/nmat856
5. Republic of Namibia Ministry of Works and Transport (2011) Namibia Rainfall performance at Specified Places
6. Shirtcliffe NJ, McHale G, Newton MI (2009) Learning from superhydrophobic plants: The use of hydrophilic areas on superhydrophobic surfaces for droplet control. Langmuir 25:14121–14128. doi:10.1021/la901557d

Chapter 7
Intelligent Housing for People and Technology: Examining Sustainable Housing Beyond the Technical and the Opportunities for Design

Phillipa Marsh

Abstract Housing is often identified as a core contributor to the UK energy concerns, with figures of 27% (Edwards and Hyett, Rough guide to sustainability, RIBA, London, 2002; Department of Energy and Climate Change, Digest of United Kingdom energy statistics, 2009) of the country's CO_2 emissions produced within the home. This concern has created a new perspective on housing design within Architecture and the Built Environment with an increasingly emphasis on technical efficiency. Intelligent solutions within housing environments are increasingly playing a sustainable role; phase change materials are now incorporated in paint and wall surfaces and reactive monitoring offers wider interactivity for homeowners. Whilst it may be easy to assume a trickle-down effect of these technologies in some form into mainstream housing, the sociological factors are also central to enabling sustainable results. In losing sight of this, many technically-oriented built environments have been significantly less efficient during their occupancy than predicted (Roy and Herring, Technovation J 27(4):194–203, 2007; Chappells and Shove, Build Res Inf 33(1):32–40, 2001). To compound this issue, much of sustainable architectural literature appears divided of their view of technologies to achieve sustainability; either overtly promoting technical solutions to solve sustainable agendas or pessimistically shows technological inclusion as a 'doom-and-gloom' scenario. This paper looks takes a more rounded open view, in examining intelligent technology beyond functional efficiency, and does not enter into the beneficial/detrimental debate of techno-centric sustainable design. This research will consider the inclusion of differing technologies within current sustainable projects. Examinations will relate the primary functions of the sustainable housing technologies alongside their social context to outline how these

P. Marsh (✉)
Department of Architecture and the Built Environment,
University of Nottingham, Nottingham, UK
e-mail: laxpm2@nottingham.ac.uk

P. Breedon (ed.), *Smart Design*,
DOI: 10.1007/978-1-4471-2975-2_7,

technologies can offer technical efficiency and wider social values. This work therefore highlights the need to explore intelligent technologies from a broader view of the overall environment and not simply secularly within a technical component or system.

Keywords Energy efficiency · Sustainability · Technology · Housing

7.1 Introducing Energy Efficient Housing and the Essence of Technology

Within the UK, housing has become a principle factor of the nation's energy concerns, with almost a third of UK carbon emissions produced from burning of fossil fuels in domestic environments [1, 2]. In light of this, continual attention has been placed on technological efficiency, promoting technological inclusions as means to achieve efficiency [3]. This dominating technocratic emphasis is argued to misinterpret sustainable philosophies, losing sight of the socio-cultural elements [4]. Equally, some technically-oriented environments can be significantly less efficient than predicted due to the limited view of the social complexity [5, 6].

This paper examines sustainable design that is focused on energy efficient technologies and the consideration of occupants of sustainable housing projects specifically. Mainstream housing present a substantial challenge to sustainable design in its relationship between society and nature. Dwellings can be seen to represent social practices that produce and transform different natures and different values [7]. Equally housing can provide a vehicle to aid in developing elements of community; creating sustainability in a resource efficient manner [4] and reflects relationships between individual, family and community [8].

The emphasis of functionalised technology clearly dominates much of sustainable housing thinking however, this view provides a restricted and basic perspective on technology; as neutralised tools. The notion of improving buildings' functions and the technical elements were at the core of sustainable efficiency. Whilst this is important, it often curbs recognition of the housing environment and the presence of its social values. Literature often shows an underlining persistence to this standpoint; neutralising technology as the key to achieving energy efficiency. In this, Feenberg's [9] instrumentalisation theory offers a useful an analytical approach. Here technology's essence is based on crosscutting the boundaries between the social and the technical, not the distinction of these ingredients, as seen in Fig. 7.1. This model works on two levels. The first level seeks to find affordances that can be mobilised in devices and systems, and the second level introduces designs that can be integrated with other already existing devices and systems that have various social considerations [10].

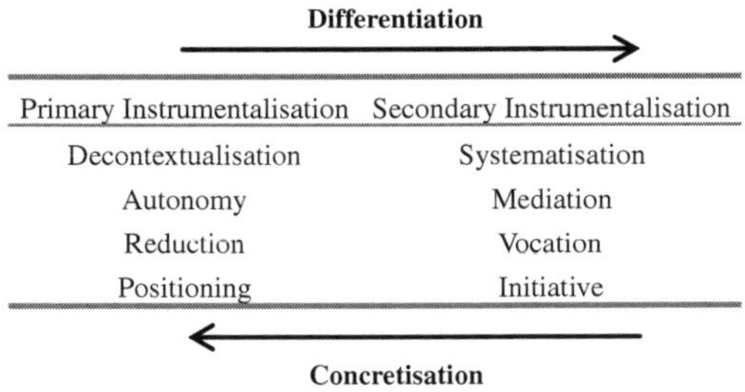

Fig. 7.1 Adapted diagram of the instrumentalisation theory [9]

The value of this approach is that it has absorbed wider grounded theories, and evolves these within a pragmatic realised context. The primary level of the model is similarly reflective of Heidegger and Habermas' differentiation theories [11, 12]. Technical and social differentiation can be overcome in the secondary level as contextualisation [13, 14]. Thus by taking a two-level perspective, the wider relationships to broader contexts can be presented with the technological perspective; how objects are integrated into natural and social environments.

The inclusion of energy efficient tools within sustainable housing enables a means of determining the situation and discovering a range of technically efficient answers [15]. Technically efficient structural components, construction details or the inclusion of appropriate appliances/devices are commonly presented in the literature as measures to achieve efficiency. These are then proven successful through measureable factors; such as statistics, ratings, external reports, or abiding with predetermined conditions in regulatory codes, thus limiting the potential socialised values. However, this appears to skim the surface of already establish perspectives where technology is often discussed as part of value-laden networks [16] or where socialised values directly shape technological systems [17, 18].

7.2 Realising the Social Within Energy Efficient Housing

Having outlined the functional dominance of technology, case study research was conducted to determine a wider view of the technological context of energy efficient housing. Two projects were chosen as pilot studies within the Nottingham location that aimed to minimising the environmental impact by maximising energy efficiency through new technologies. However, it is important to note that as pilot studies, results are not fully conclusive but do provide an insight into users' actual experiences of existing energy efficient housing.

Results from both interviews highlighted aspects where occupants coped with technological difficulties, relative to unfamiliar technologies and unfamiliar user actions. Whilst evidence of similar results has been found in other research projects, the impact of how users felt; their experiences and the subsequent perceptions of the technologies, are less covered. The process of persevering and whether users were happy with the results of their actions offers a link to the potential for accepting the technology into their social environment. Perseverance was discussed in terms of their involvement and the resulting rewards they receive; whether their actions delivered positive or negative results and if these were considered worthwhile.

Occupants often showed that by physically noticing the results of their actions, they reflected positively on the technology. There is an additional value that the user notes; in working out the mode of operation successfully. However, not all evolutionary perceptions of the technology resulted in a positive reflection. Some technologies were viewed negatively after users persevered with unfamiliar technology and their actions. The users' understanding of the technologies' functions were shown to impact on their awareness of wider conditions. Users appeared more aware of their affect on efficiency. As they became familiar with the technology, they showed awareness not simply on how the technology works but in some cases why it works to achieve efficiency.

These results are relative to Norman's theory on technological attachment [19], where users' behavioural involvement in operating the technology is a key level to developing an attachment and acceptance of that technology being embedded in context. These values form user's attachment to a technology and can be short/long term values dependant on the attained stage [19]. Forming attachment is often shown in line to the concept of self [20, 21] thus repeating the social dimension. Once something is accepted, it often belongs to one of the three intermediate stages of attachment: early and mature ownership and predisposal. This, in part, is resonant to notions of lifecycle, and often referenced in response to retaining attachment in alternative views of sustainability. Recommendations to prolonging lifespan through durable or attached values, in essence, relates to the sustainable premise, but one which again appears restrictive in much of sustainable housing discussions [20, 22]. The notions of technological acceptance parallel to wider literature that considers technological efficiency with housing. Rybczynski illustrated that efficiency is more in tune with user comforts, demonstrating that inefficient used or misused of technologies often occurred because the technology addressed the home physical, behavioural and psychological comforts of the occupants [23]. Harris echoes this idea, showing the impact of comfort on the adoption of cast iron stove technology in a historically US housing context [24]. His work stresses that to be accepted into daily life, the technology needed to offer values of comfort in terms of convenient efficiency.

From this, a technology can be positively and negative received into a social context. It is not simply the answer to say that a technology functions in one way and then to evaluate how the user interpret this. More so, these findings shows that understanding the contextualised embedding of technologies within an environment can relate to both users' actions, involvement and the processes they experience as well as the technicalised functions of the technology. The varying results from all of these can therefore lend to the mediation of technology.

Interestingly within much of the established sustainable housing literature, notions of domestic values and meanings are limited reference, if not ignored. This research has found that in these sustainable case studies, housing values are still present in occupants' considerations. In both cases, the house appeared to offer a distinct value to the occupants primarily as a means of energy efficiency use rather than as domesticated home: "The character of the house is different. Because of the technology and what you have to do, it feels like more of a technical device than a home."

Whilst these results reflect technised values, in part due to the projects' premise as conceptual exemplars, the findings also suggest notions of a different value to these house environments. In some cases occupants discussed how the original design offered the potential for homeliness. However, others discussed the difficulties that these designs had to fully form a value of home. The traditional notion of home appears in these environments by the occupants; in adapting and developing the environment to suit their requirements and tastes. The designs restrictions appeared to restrict users' ability to adapt to their preference. Whilst these are physical adaptations, the premise appeared as a desire to personalise.

Occupants suggested a need to make a home theirs to which reflected the need to physical change the design. This corresponded with perceived missing features, focal points and valued features that provide a feeling homely comfort. In both cases the fireplace was outlined as a key homely feature, where it offers a means of experiencing the home through values of comfort and intimacy. The experience within the home relates to developing values and associated meanings that merge the space to the individual involved. These ideas ultimately return to Feenberg's paradigm of the house environment framed by valued social meanings and technological inclusion [9]. Housing is therefore not simply about the physical function as a shelter but as psychological security, drawing connotations of societal status, a communal or family base and a haven for privacy to maintain well-being [25]. Danby addresses similar thinking in context to design, suggesting that housing can reflect the link between individual, family and community [26], thus a home has broader connotations than the functional house referenced.

From these results, this work suggests that it is not simply enough for post-occupancy evaluation to consider technologies and users efficient use or misuse. More so this research has shown that there is a different level of understanding to technology when implemented into the social and natural environment; understanding the mediation or embedding into the social context. This mediation, outlined by Feenberg's realisation stage [9], has been shown to be lacking in the current perspectives of energy efficient technologies within housing. However, as these results show, a fuller concretised understanding of technology can be determined by embedding a social view within this more intelligent environment.

7.3 A 'Social' Conclusion Looking Beyond the Technical

From this paper, an apparent limitation to the socio-technical perspectives has been presented. Not only is there a need to consider the social elements in context to efficiency within sustainable housing but also to consider the interrelating factors between technology and social contexts within the housing environment. In understanding the technological emphasis within sustainable housing literature, there is a clear socio-technical disparity of extremes. To one extent, sustainable design literature appears to separate the technology from the social, considering technical functions to be a means of measurable efficiency. To the other, the social dimension was included but from a limited functionalised physical view; evaluating occupants' use of the technology. By emphasising one over others appears to cloud the picture; offering an understanding of the technical environment, but not reflecting the social constructs that influence this network.

These critical discussions provide opportunities for design to reconsider the domestic environment, and to reinterpretation current architectural views on energy efficient housing, as a means of reconnecting to the domestic environment. This work is not simply showing that the social dimension is important and needs wider consideration. Discussions support the need to reinterpret energy efficient housing environments beyond the view of technical efficiency and expand considerations to social notions of home, dwelling, technology, comfort and efficiency. Through a wider understanding, sustainable practice has the opportunity to reconnect with the domestic environment alongside the application of technical efficiency and developing a broader view of sustainable housing; in the perception of efficiency alongside the values of the domestic environment.

Acknowledgments This paper acknowledges the help and guidance from Graham Farmer of the University of Newcastle and Michael Stacey of the University of Nottingham.

References

1. Edwards B, Hyett P (2002) Rough guide to sustainability. RIBA, London
2. Department of Energy and Climate Change (2009) Digest of United Kingdom energy statistics. DECC
3. Smith A (2003) Transforming technological regimes for sustainable development: a role for alternative technology niches? Sci Public Policy 30(2):127–135
4. Guy S, Moore S (2005) Sustainable architectures: cultures and natures in Europe and North America. Taylor & Francis, London
5. Rochracher H, Ornetzer O (2002) Green buildings in context: improving social learning processes between users and producers. Built Environ J 28(1):73–84
6. Guy S, Shove E (2000) A sociology of energy, buildings and the environment. Routledge, London
7. Macnagton P, Urry J (1998) Contested natures. Sage Publishers, London
8. Mallett S (2004) Understanding home. Sociological Rev 52(1):62–89
9. Feenberg A (1999) Questioning technology. Routledge, London
10. Feenberg A (2005) Critical theory of technology. Tailoring Biotechnol 1(1):47–64
11. Heidegger M (1977) The Question concerning Technology (trans: Lovitt W). Harper Row, New York
12. Habermas J (1970) Technology and science as 'Ideology', in towards a rational society. Beacon Press, Boston
13. Veak T (2006) Democratizing technology. Suny Press, New York

14. Feenberg A, Feng P (2008) Thinking about design: critical theory of technology and the design process. In: Vermaas PE et al (eds) Philosophy and design. Springer, Berlin
15. Williamson T, Radford A, Bennetts H (2003) Understanding sustainable architecture. Taylor & Francis, London
16. Latour B (1993) We have never been modern. Harvester Wheatsheaf, New York
17. Bijker W, Hugher T, Pinch T (1989) The social construction of technological systems: New directions in the sociology and history of technology. MIT Press, Cambridge
18. MacKenzie D, Wajcman J (1999) The social shaping of technology. OUP, Oxford
19. Norman D (2004) Emotional design. Basic Books, New York
20. Ball AD, Tasaki LH (1992) The role and measurement of attachment in consumer behaviour. J Consum Psychol 1(2):155–172
21. Kleine SS, Kleine RE, Allen CT (1995) How is a possession "me" or "not me"? J Consum Res 22(3):327–343
22. Till J (2009) Architecture depends. MIT Press, Cambridge
23. Rybczynski W (1987) Home: a short history of an Idea. Penguin Books Ltd, Baltimore
24. Harris H (2010) Conquering winter: US consumers and the cast-iron stove. In: Shove E, Chappells H, Lutzenhiser L (eds) Comfort in a low carbon society. Routledge, London
25. Daly G, Daly G (1996) Homeless: policies, strategies, and lives on the street. Routledge, London
26. Danby M (1993) Privacy as a culturally related factor in built form. In: Framer & Louv (1993) Companion of contemporary architectural thought, Routledge, London

Chapter 8
Interrogating Hyperfunctionality

Sarah Kettley

Abstract It is proposed that one of the central visions of smart design is to extend the fundamental working philosophies of *fitness for purpose* and *the disappearing tool* as found in product design and human computer interaction. This 'hyperfunctionality' may be achieved spatially, through distributed networks, and/or temporally, through reactive materials and adaptive algorithms. There is a healthy history of criticism of products which can fit a need so well as to design out the creativity of the people using them, or which are so seamless as to be beyond either appreciation or skepticism, and as design becomes exponentially 'smarter', we will be wise to revisit these arguments regularly. This paper presents the case for shifting the goals of design away from creating new consumer 'needs', and argues for a design methodology based on Umberto Eco's concept of the *open work*. The open work supports multiple associative paths formed by viewers through art objects within different historical eras, or from different cultural perspectives, and is suggested here as a useful way of interrogating the working philosophies of Smart Design and as a potential route to designing what Illich called *convivial tools* (Illich, Tools for conviviality. Calder and Boyars, London, 1973).

Keywords Open design · Conviviality · User creativity · Hyperfunctionality · Authorship

S. Kettley (✉)
Nottingham Trent University, Nottingham, UK
e-mail: sarah.kettley@ntu.ac.uk

P. Breedon (ed.), *Smart Design*,
DOI: 10.1007/978-1-4471-2975-2_8, © Springer-Verlag London 2012

8.1 Hyperfunctionality and Smart Design

The critical position presented in this paper is based on a model of Smart Design that extends the existing design trajectories of *fitness for purpose* and the *disappearing computer*. It is suggested that one of the key characteristics and goals of Smart Design is the seamless conjunction of these two functional principles, resulting in what is understood to be an optimal relationship between user and product. For the sake of argument, I have called this idealized combination of use-value and perceptual qualities 'hyperfunctionality'. This section briefly introduces these two contributing principles and gives examples of hyperfunctionality in Smart Design today.

Fitness for purpose works on the assumption that the user is knowable by the designer and that problems have answers, which only need to be unearthed by the work of design: designers are habitually engaged in "changing existing situations into preferred ones" [1]. Fitness for purpose can be taken to extremes in the personalisation of design for individual users and contexts, supported today by advances in technologies and marketing to create mass customisation based on individual purchasing preferences. Here at NTU, undergraduate product design students are taught to synthesise "a wide range of factors from the understanding of consumer and market needs", and are encouraged to "produce challenging and inventive solutions to a wide range of design problems and briefs". Students should be able to define problems, design solutions according to user needs, and "*ensure fitness for purpose*" [2]. Design consultancies such as IDEO propose that by observing the everyday creative solutions of people (potential customers), problems or needs are revealed, and new opportunities for design can be identified.

While such predetermination of needs sets up useful criteria for the measurable evaluation of design proposals, it also embodies a pervasive philosophy of progress and convenience, described with reference to the Greek myth by both Bruno Latour and Ivan Illich as Promethean [3, 4]. The capitalist rewards to be had in fitting a product effectively to a user exacerbate the power relationship created between the naïve user and the knowledgeable designer. In one book used widely in design pedagogy, an 'expert voice' explains the role of research in helping a "large corporate client" to "tap into the endless opportunities" presented by developing nations, detailing how a range of techniques led to "a deep understanding of the diverse cultures being examined…"[5]. In many cases, such user centered design is based on entirely laudable intentions, to help others, but even when the commercial imperative is not laid as bare as in this example, this, like philanthropy, can and must embody an unequal relationship. As Illich said, this has been, and continues to be, an age in which human needs are "shaped by professional design" [6].

The notion of *disappearance* in design is not entirely new. Peter Behrens, working for AEG, grappled with a "transformation of aesthetic notions" to design the effective electrification of the home in the early 1900s, and "if the rapid electrification of everyday life just three generations ago is any guide, embedded computing will not prove controversial" [7]. Electrical motors disappeared from view, encased in the streamlined packages of car and train bodies, lamp housings, and the protective casings of meat slicers and other equipment with moving parts, all designed to express abstract qualities of modernity, and making full use of the new materials and manufacturing processes embodying modern progress, such as aluminium, rubbers and plastic. Then, the commercial imperative was also strong, with the new field of industrial design charged with manipulating consumers' desires and creating irresistible products to streamline consumption as much as the design aesthetic itself [8]. We now find ourselves in the similar position of having to review aesthetic design principles in order to create meaningful products, this time with the components of pervasive computing, which lack form and expression, just as electricity once did [9].

Products can disappear in more than one sense: physically and perceptually (phenomenologically). Moore's Law and intangible qualities combine in materials like electricity and computing to challenge the designer more accustomed to handling materials directly. These forces cannot be manipulated in the same way metals and plastics might be, we can only work on them indirectly, through the components that carry their charge, or through those materials that make apparent their effects—heat, light, movement and sound for example. Daniel Weil's Radio In A Bag [10] demonstrated the problem graphically by deconstructing the parts of a commercially available radio and arranging them in a clear plastic bag, visible and yet unknowable. The components are revealed for all to see, but their meaning is all the more mysterious, and they tell us nothing about the material of radio waves or electricity [11]. With advances in ubiquitous computing, based on peer-to-peer networks of many small computational nodes, developed by such research projects as Smart Dust and Speckled Computing, sensing, processing and actuation are hidden inside familiar objects, or combined with substrate materials to create new malleable computational compounds.

Phenomenological disappearance entails designing products as tools which achieve optimal fitness for purpose, as above, and which work so effectively as to be unnoticeable in use—hence the windows paradigm—we focus on what is being achieved through the transparency of the technology, rather than on the material of the technology. The tool re-presents itself to our attention when it does not work seamlessly. This paper aims to contribute to the discourse challenging the lazy logic, which suggests that the converse must also be true; that a tool that presents itself to our attention must be defective.

Hyperfunctionality then, is a combination of fitness for purpose (seamless integration with the needs of a user) and disappearance (physical and/or perceptual). As with any paradigm, the utopian ideals driving Smart Design are not neutral, and calls for a new community of practice in ubiquitous computing user experience [12] need to be engaged with in as thoughtful a way as possible if we are not to merely reproduce design "in the service of an increased servitude of people to commodities" [3]. As a word of warning, consider the implicit interaction model of ubiquitous computing as outlined by Stefan Poslad [13] in which people are 'bottlenecks' in complex systems, humans cannot understand machine interaction, human cognitive and haptic capabilities are likely to be overloaded, and systems must be automated as far as possible.

8.2 Convivial Tools: A Critique of Hyperfunctional Design

It is hard to critique user centred design, because it has such well meaning and egalitarian aims; it is the unquestioning, or worse, cynical adoption of such aims for commercial gain which need to be repeatedly challenged. Ivan Illich wrote in the 1970s on the positivist and often destructive nature of institutionalized professions and capitalist infrastructures. He proposed in contrast, a socialist model of design, which would support "autonomous and creative intercourse among persons and the intercourse of persons with their environment" [14]. In particular he was at pains to differentiate the status of the trapped consumer from that of the free man, emphasizing the need people have not only to obtain things, but to act upon their world. As a methodology for achieving this, Illich called for "negative design criteria for technological devices" [15] in order to put constraints on the institutionalization of 'needs' and the fitting of production to monetary gain over individual human agency. Obviously, the line between such an idealist approach and dictatorship can be very fine yet we can all recognize how tools have come to demand purpose and expertise from us: we feel inadequate when we cannot program our smart televisions effectively, or when the automated air conditioning system appears to take on a will of its own [16].

Despite the radical nature of the socialism espoused by Illich, his ideas can increasingly be found worked through in Human Computer Interaction and Interaction Design. Jaron Lanier, one of the first pioneers of virtual reality, is concerned with the ways in which small changes to a design can alter the behaviour patterns and perceptions of identity of its users. He holds that working with information technology is effectively an engagement with social engineering, and calls for greater philosophical understanding in the profession [17]. Like Illich (and others he lists before him), he is concerned with the impacts of technology on the distribution of wealth and the erosion of the individual [17]. Elizabeth Sanders writes about creating scaffolds for everyday creativity, rather than providing people with pat solutions to everyday problems [18]; also in the 1970s, Victor Papanek was espousing Design for Need and developing ideas on the role of design in what he and others called 'countries of the periphery', or the Third World. In so doing, he adopted what Sanders might agree was a scaffolding approach to form and aesthetics—creating the functional object with reclaimed materials, to be combined and decorated by the user in the cultural context of use. Alison J. Clarke's essay illustrates this with an image of a Tin Can Radio before and after embellishment with felt and seashells by a user in Indonesia [19], and in fact this example illustrates Baudrillard's differentiation between use and ownership [20]. Dunne and Raby, among other Critical Designers, create technology as critique in order to make the social and ethical implications of technology graspable and therefore open to discussion [21]. The problem for Critical Design at the moment is that it remains close to an art practice, especially in its framing in the gallery space, with some wondering what impact it can have on real world design, which persists in operating in the name of opportunism [22, 23]. *Graspable* and *palpable* media have in turn become the focus of a number of research programs, where the terms are used to denote both a physical tangibility supporting cooperative design processes [24], and a cognitive availability: 'noticeable' and 'understandable' [25, 26]. Corporations need to be aware of how they use these terms: Phillips says graspable interfaces give people the 'feeling' of being control, while proposing an ambient intelligent living space. The emphasis remains with the institution *providing* personalization and emotion rather than creating a context for it [27]. Many more designers work in variations of an Action Research framework, seeking to expand the field of design to include large, complex social, political and geological issues, which Latour has challenged us to see as the real 'matters of concern' for design [28]. In his keynote address to the Networks of Design conference in 2008, Latour described design as in thrall to the idea of progress for its own sake, and suggested that its "heroic, Promethean, hubristic dream of action" be tempered by the "modesty, care, precautions, skills, crafts, meanings, attention to details, careful conservations, redesign..." of, among other things, craft. As he put it, "we have to be radically careful, or carefully radical" [28]. This call to action comes at an excellent time, as craft throws off a somewhat pervasive despondency regarding its relationship with technology, to rediscover its far more productive and interesting role in philosophy. Speaking at a Crafts Council 'Craft Rally', Jane Webb recently showed how craft has stood apart from art and design in its continuing emphasis on the relationship between maker and material, a model that Latour, Alfred Gell and Tim Ingold have all drawn on in developing concepts of the thing and the object [29]. The author has also been engaged in developing a model of craft that demonstrates its parallels with the philosophical concept of authenticity as it has changed over the centuries [30–33]. While these publications have presented the rationale and protocols for craft as a crucial set of practices for critical awareness in the production and consumption of products, this paper suggests an 'open' design strategy as just one approach.

8.3 Designing for Conviviality: The Open Work

Design as a function of mass manufacture and mass consumption is dependent for sales on clear marketing messages. However, the clearer the message, the less manouevre there is for creative meaning making on the part of the consumer. Eco draws a comparison with chemical reactions, and the possible paths of molecules within a system [34]. He suggests that in trying to model the behaviour of any given molecule in the system, we introduce order to the system, closing down potential paths. Meaning may be read *as given* in the simple message, or actively constructed when faced with a more complex system of possibilities. Eco goes on to discuss how contemporary art operates to open up paths for meaning through an interplay of the known and unknown, constituting the "very essence of the open work" through "an oscillation between rejection and preservation". In this way, 'noise' is managed, while the challenge to the viewer to create meaning is maintained.

User centred design, when deployed without imagination, forgoes this poeticism, as it proscribes against the author's voice—Plain English is intended to be understandable to as many people as possible, but the cost is high. What if the goals of design are rewritten to champion user experience, intelligence and creativity? What if we intend the user to be critically aware of her role, cast as consumer, as needful of our technology, and as essential to the larger concepts of modern progress? Then 'need' is recast as open, a deepening of interaction and communication rather than of repetitive consumption: "Think of the multiple uses for a simple tin opener, whereas an electric one, if it works at all, opens only some kinds of tins, and costs one hundred times as much" [35]. Affordance, once the epitome of good design, embodies narrow meanings in its quest for clarity, as instructions "for" or "about ongoing life" become built into the form of the product itself: "sometimes these instructions are attached to the item and must be read. In more thoroughly designed goods, the shape, colour, and provoked associations speak to the user about the way the item must be handled" [36].

In supporting people's ability to "endow the world with their personal meaning" [37] designers must also reinstate themselves as authors, and aim to design poetic products. In terms of visual language, this may mean a fundamental rethink for much of design education, which has come to rely heavily on established visual languages, and which is certainly challenged by young students' preoccupations with commercial success as epitomized by today's generic forms of consumer electronics.

8.4 Towards an Open Design Methodology

The open product poses a fundamental challenge for design, at odds with its familiar goals of systematization and optimization. Alice Rawsthorn gives an early example of the benefits of standardization of manufacture in the arrows of the Chinese Emperor Qin Shihuangdi in the 200s B.C., who found his soldiers unable to fit fallen men's arrows to their hand made bows [38]. But we should also be aware of the ways in which such systems of construction come to determine our worlds, as loom widths determine woven cloths, which in turn determine living spaces, and the girth of a horse determines plough widths and eventually the standardized gauge of the railway track [39].

Eco speaks of an oscillation between the known and the new in the open work—a balance of defining or predetermining and opening up or problematizing. Examples of such design work can be found: Critical Design seeks to create a distance between the user and the object in order to create a context for questions; Gaver et al. have sparked much interest in the notion of ambiguity and playfulness in design [40]; and Niederrer proposes the 'performative object', which acts to destabilize expectations and promote social interaction in use, as in her 'Social Cups' [41]. Yolande Sengers in her paper *Too Smart*? unravels users' perceptions of an automated air conditioning system, finding that it either turns the user into a resource manager or removes agency totally from them; she suggests that in smart design, we all too often start by asking a technical question grounded in resource management instead of asking about everyday practice first [16]. In Contemporary Craft, makers actively avoid received visual language, instead working through visual and material research strategies to build new formal languages [42]. Some genres of craft have in fact been criticized for an over emphasis on formal concerns and a turning away from the social [43] but this might yet offer an approach for design through a return to simple creative activities such as drawing, manipulating materials, and gathering of resources in the creative space, allowing an internalization of the familiar to ferment over time and return, metamorphosed 'through the hands' as something new. In the evaluation of the author's *Friendship Jewellery* (a handbuilt suite of networked 'smart' neckpieces revealing identity and proximity), it was apparent that the participating user group was able to shift between useful and contemplative configurations for the work; they were surprisingly adept at managing the visibility of the world around them. In this, they were to a great extent in control of the status of the jewellery as things (open to new readings) or objects (embodying reified meanings), and were even able to combine these statuses in a sophisticated simultaneity [30]. I believe this demonstrates the great creativity of the women involved and gives the lie to the way users are often implicitly written as passive receivers of meaning.

Design has a close relationship with linguistics and literary theory: it was the use of visual metaphors that underpinned the development of the desktop interface and the personal computer, after all. Donald Schön taught us about conversations with the situation at hand [44], and Krippendorf continues to discuss the evolving relationship between technology-centered and human-centered design [45, 46]. As HCI, Interaction Design and Smart Design reconsider user-centered methodologies and techniques, and designers are increasingly recast as facilitators [47], we should bear in mind the equivalent, and problematic, role of the author in literary theory: Barthes, writing in 1977, proposed that in the very act of writing, the author necessarily dies, giving way entirely to the destination of the text with the reader [48]. The identity of the author is completely negated in the process. However, a wonderful reaffirmation of the role can be found in Lanier's thesis that "you have to be somebody before you can share yourself" [49]. This echoes what investigations into craft practice have revealed—the importance of the role of the maker as author, as a recognizable identity growing and taking responsibility for her actions and utterances. The disappearance of the designer in the industrial system is a central problem for the production of tools for conviviality, and designers need to work to maintain their identity in the design and manufacturing process. Authorship is convivial.

References

1. Simon H (1996) The sciences of the artificial. MIT Press, Cambridge
2. Nottingham Trent University (2010) BA product design program specification. pp 1–2
3. Illich I et al (2010) Disabling professions. Marion Boyars, London

4. Latour B (2008) A cautious Prometheus? A few steps toward a philosophy of design. In: Networks of design international conference, Falmouth University, 3–6 Sept 2008
5. Visocky O'Grady J, Visocky O'grady K (2006) A designer's research manual: succeed in design by knowing your clients and what they really need. Rockport, Beverly, p 78
6. Illich I et al (2010) Disabling professions. Marion Boyars, London, p 14
7. Thackara J (2006) In the bubble: designing in a complex world. MIT Press, Cambridge, p 199
8. Maffei N (2003) The search for an American design aesthetic: from art deco to streamlining. In: Benton C, Benton T, Wood G (eds) Art Deco 1910–1939. V & A, London, pp 361–369
9. Thackara J (2006) In the bubble: designing in a complex world. MIT Press, Cambridge
10. Weil D (1981) Radio in a bag. V&A (n/d) search the collections. http://collections.vam.ac.uk/item/O85208/radio-radio-in-a-bag//. Accessed 6th March 2012
11. Dunne A (2006) Hertzian tales: electronic products, aesthetic experience, and critical design. MIT Press, Cambridge
12. Kuniavsky M (2010) Smart things: ubiquitous computing user experience design. Morgan Kaufmann, Burlington
13. Poslad S (2009) Ubiquitous computing: smart devices, environments and interactions. Wiley, Chichester
14. Illich I (1973) Tools for conviviality. Calder and Boyars, London, p 24
15. Illich I (1973) Tools for conviviality. Calder and Boyars, London, p 31
16. Sengers Y (2011) Too smart? In: Entanglements, symposium of the OPEN design research society special interest group, Nottingham Trent University, April 2011
17. Lanier J (2010) You are not a gadget. Penguin Books, London
18. Sanders L (2006) Scaffolds for building everyday creativity. In: Frascara J (ed) Design for effective communications: creating contexts for clarity and meaning. Allworth Press, New York
19. Clarke AJ (2011) The anthropological object in design: from victor Papanek to Superstudio. In: Clarke AJ (ed) Design anthropology: object culture in the 21st Century. Springer, New York, pp 74–87
20. Baudrillard J (1968) (J Benedict Trans) The system of objects. Verso, London
21. Moggridge B (2007) Designing interactions. MIT Press, Cambridge
22. Malpass M (2009) Contextualising critical design: a classification of critical design practices. In: Design Connexity proceedings of EAD conference 09, Robert Gordon University, Aberdeen
23. Hunt J (2011) Prototyping the social: temporality and speculative futures at the intersection of design and culture. In: Clarke AJ (ed) Design anthropology: object culture in the 21st Century. Springer, New York, pp 33–44
24. Hornecker E (2002) Understanding the benefits of graspable interfaces for cooperative use. In: Proceedings of Coop'2002 (fifth international conference on the design of cooperative systems) in Saint Raphael, IOS Press, 4–7 June 2002, pp 71–87
25. Palcom (2008) Making computing palpable. At http://www.ist-palcom.org/. Retrieved 8 Aug 2011
26. Pollini A (2008) Experimenting with an Ubiquitous computing open architecture. Phd Thesis, University of Siena. Available at: http://www.ist-palcom.org/publications/files/PhD-Thesis-Pollini.pdf. Accessed 8 Aug 2011
27. Van Loenen EJ (2003) On the role of graspable objects in the ambient intelligence paradigm. Media interaction group Philips research labs, Eindhoven. In: Proceedings of sOc'2003 (Smart Objects Conference), Grenoble, 2 May 2003
28. Latour B (2008) A cautious Prometheus? A few steps toward a philosophy of design. In: Networks of design international conference, Falmouth University, 3–6 Sept 2008, p 7
29. Webb J (2010) The craft of enlightenment. In: Manchester Craft Rally, Manchester Metropolitan University, 8 Dec 2010
30. Kettley S (2008) Crafting the wearable computer: design process and user experience. Edinburgh Napier University, Edinburgh
31. Kettley S, Smyth M (2004) The materiality of wearable computers—craft and authentic user experience. Des J 7(2):32–41

32. Kettley S (2008b) Parallels in the evolution of craft and authenticity. In: Crafticulation and education, international conference of craft science and craft education, University of Helsinki, Helsinki, Finland, 24–26 Sept 2008
33. Kettley S (2010) Fluidity in craft and authenticity. Interactions 5:12–15
34. Eco U (1989) The open work. Harvard University Press, Cambridge
35. Illich I (1973) Tools for conviviality. Calder and Boyars, London, p 34
36. Illich I (1973) Tools for conviviality. Calder and Boyars, London, p 73
37. Illich I (1973) Tools for conviviality. Calder and Boyars, London, p 75
38. Rawsthorn A (2008) What is good design? The New York Times, Friday, June 6
39. Pye D (1968) The nature and art of workmanship. Design Handbooks, London
40. Gaver W, Beaver J, Benford S (2003) Ambiguity as a resource for design. In: CHI'03 proceedings of the SIGCHI conference on human factors in computing systems
41. Niederrer K (2007) Designing mindful interaction: the category of performative object. Design Issues, 23(1)
42. Kettley S (2005) Crafts Praxis as a design resource. In: Engineering and product design education conference, crossing design boundaries, Edinburgh Napier University, Taylor and Francis Group, London, Sept 2005 pp 545–549
43. Cohn S (2007) Cyber-jewels and techno-gadgets; smart ornament. In: Filter issue 64, summer 2007: ReSkin—The future of wearable technology. Adelaide: Australian network for the arts and technology, pp 8–10
44. Schön D (1983) The reflective practitioner. Basic Books, Cambridge
45. Krippendorf K (2005) The semantic turn: a new foundation for design. CRC Press, Boca Raton
46. Krippendorf K (2011) Principles of design and a trajectory of artificiality. J Prod Innov Manage 28(3):411–418
47. Siu KWM (2003) Users' creative responses and designers' roles. Des Issues 19(2):64–73
48. Barthes R (1977) The death of the author. In: The death of the author. Image, music, text, Fontana, London, pp 142–148
49. Lanier J (2010) You are not a gadget. Penguin Books, London, p xiii
50. Illich I (1973) Tools for conviviality. Calder and Boyars, London

Chapter 9
Semefulness: Touch, Design and Meaning

Anne Cranny-Francis

Abstract This paper explores the multiple significances (semefulness) of touch, as experienced by us as embodied subjects. Prompted by the development of a range of touch-based technologies and by new writings on the senses, I explore meanings of touch—connection, engagement, contiguity, differentiation, positioning—for their contribution to our understanding of the world and of our own embodied subjectivity.

Keywords Touch · Tactility · Technology · Connection · Engagement · Contiguity · Differentiation · Positioning · Body · Embodiment · Emotion · Knowledge · Culture

9.1 Introduction

Touching might be described as the state whereby two entities or objects are so close that no space remains between their boundaries or surfaces; or as the process whereby a message from the brain is sent to the muscles of the hand to position itself beside or around an object or entity so that neural receptors in the skin are able to send a message back to the brain that this object or entity can be physically sensed (as on the MIT Touch Lab website [1]). This is the physical practice and sensation of touch but this study addresses more than this; it is concerned with what that physical sensation *means* to individual human subjects.

A. Cranny-Francis (✉)
University of Technology Sydney, Sydney, Australia
e-mail: anne.cranny-francis@uts.edu.au

P. Breedon (ed.), *Smart Design*,
DOI: 10.1007/978-1-4471-2975-2_9,

Erin Manning notes that when she started writing her book, *The Politics of Touch: Sense, Movement, Sovereignty* "finding work on the senses was such a challenge that I welcomed anything I could lay my hands on" [2]. Since then we have seen a proliferation of writing about the senses [3–9], and some specifically on the sense of touch [10, 11]. This paper explores the experience of touch as sign (*seme*) and sensation, an embodied practice that is both culturally and socially specific and which locates us in the world in relation to other beings and objects. Accordingly, the deployment of touch reveals the nature of both embodied subjects and the society and cultures in which they live and in this sense is *seme-ful*—multiply significant, physically, emotionally, intellectually, spiritually, politically. Given the current proliferation of touch-based technologies—both technologies we touch such as haptic technologies and touch screens, and textile-based technologies that touch us—it is more than ever important to explore this semefulness.

9.2 Touch

Constance Classen's anthology, *The Book of Touch* is a multi-disciplinary collection, featuring work from anthropology, healthcare, aesthetics, art, dance, and literary studies. Classen writes:

> Touch is not just a private act. It is a fundamental medium for the expression, experience and contestation of social values and hierarchies. The culture of touch involves all of culture [12].

She continues:

> One of the ideological barriers to writing about touch in culture is the customary Western emphasis on the brute physicality of touch. The sense of touch, like the body in general, has been positioned in opposition to the intellect, and assumed to be merely the subject of mindless pleasures and pains [12].

For this reason, the critical analysis of touch only developed after feminist, poststructuralist, and postcolonial critiques of the mind/body dichotomy demonstrated the suppression of 'body' within this formulation. Deconstructing this opposition meant that it became necessary to acknowledge and to analyse the role played by the senses in the formation of subjectivity and of knowing—not as a mechanical function controlled (processed) by the mind, but as an integral part of our negotiation of subjectivity and of our production of knowledge.

On 16–17 May, 2008 the Courtauld Institute in London held its first conference on *Sculpture and Touch* in order "to introduce a new impetus to the discussion of the relationship between touch and sculpture by setting up a dialogue between art historians and individuals with fresh insights working in disciplines beyond art history" [13] Just a week earlier (7 May 2008) the University for the Creative Arts had hosted the conference and exhibition, Mem*ory and Touch: an exploration of textural communication* to explore the relationship between touch and design, noting on their website the need to explore "that level of communication achieved through attention to the senses, for "to touch is also to be touched". And as such always creates a dialogue, a communication both before and beyond text." This description captures a critical aspect of the sense of touch; that it is at once a *touching* and a *being touched.*

For philosophers such as Michel Serres the Janus nature of touch is the source of consciousness:

> I touch one of my lips with my middle finger. Consciousness resides in this contact. I begin to examine it. It is often hidden in a fold of tissue, lip against lip, tongue against palate, teeth touching teeth, closed eyelids, contracted sphincters, a hand clenched into a fist, fingers pressed against each other, the back of one thigh crossed over the front of the other, or one foot resting on the other. … skin on skin becomes conscious, as does skin on mucus membrane and mucus membrane on itself. Without this folding, without the contact of the self on itself, there would truly be no internal self, no body properly speaking, coenesthesia even less so, no real image of the body; we would live without consciousness; slippery smooth and on the point of fading away [14].

We know ourselves and the world through the sense of touch, crucially including our ability to touch ourselves and *to make sense/meaning of that touch.* At the point of touch, of contact (*com-* "together" + *tangere* "to touch"), we know both the self and the other, including the other that is also the self; that can reflect on and position the self. This is a point of connection, at which we perceive connection only through the perception of difference; otherwise all would be 'slippery smooth' continuous—the world and the self as undifferentiated. In differentiating the other from ourselves, we are able also to connect knowingly with that other.

9.3 Connection

One of the fundamental properties of touch, as the Midas myth reveals [15], is that it creates a connection between individuals and things or other individuals. This connection is culturally determined or inflected—a function of distinctions created by gender, sex, class, ethnicity, religion, age, disability and so on. For example, in the 'alpha male' world of business it is considered very disadvantageous for a man (particularly) not to shake hands firmly, in a strong, powerful (stereotypically manly) way, though a man who grips hands too firmly so that he causes pain is considered to be compensating for other weaknesses—of character or intellect [16, 17]. Also, the touching between men and women in most cultures is finely regulated by parameters such as family relationship, work relationship, friendship, and familiarity. In western cultures, men and women touch only in the most formal way (e.g. handshake) unless they share a friendship or family connection. Those who touch too freely are regarded negatively because their touch signifies a closeness of connection that has not been established formally. And when close touching is necessitated by crowd behaviour (travel on a crowded train, for example), people conventionally pretend that the touch is not occurring; they mentally withdraw their consciousness from the touch so that they do not have to acknowledge the connection. These meanings are specific to particular societies and cultures. So touch also connects together a culture or a society: that is, the tactile regime of a society or a culture identifies that culture and the individuals participating within it. Equally, one of the first things people need to learn in order to live comfortably within a society or culture different from that in which they grew up is its tactile regime; the consequence of failure is not only that one is rejected as alien (connection denied) but one may also seriously offend other members.

Equally compelling is the connection established by touch between humans and things or objects. Consider this description by David Attenborough of a 1.8 million year-old stone chopping-tool discovered by anthropologist, Richard Leakey:

> Picking it up, your first reaction is it's very heavy, and if it's heavy of course it gives power behind your blow. The second is that it fits without any compromise into the palm of the hand, and in a position where there is a sharp edge running from my forefinger to my wrist. So I have in my hand now a sharp knife. And what is more, it's got a bulge on it so I can get a firm grip on the edge which has been chipped specially, which is sharp... I could perfectly effectively cut meat with this [18].

British Museum Director, Neil MacGregor notes in Attenborough's description "something of the excitement that [discoverer, Richard] Leakey must have felt" [18]. We all know this same excitement, where the touch of an object from the past—precious to our culture or society, or even just to our own family history—links or connects us to that past.

This connection is sometimes less comfortable when it applies to machines and particularly to the newest forms of technology. For example, the perceived de-skilling and enslaving of human workers to the factory machines of the nineteenth century was expressed in their description as 'hands' (of the machine). This unease is configured in the cyborg—the monstrous progeny of the 'unnatural' merging (connection) of human and machine, which has been used in science fiction since the early nineteenth century to explore the effect on human societies of the development of advanced technologies and their incorporation into everyday life. The other source of unease comes from the question of who controls the technology with which we interact, whether this is the modern equivalent of the 19th century factory owner, the modern commercial entrepreneur, or government agencies. In each case the concern is that users are being incorporated into a technological entity or assemblage of which they may have limited knowledge and understanding. For this reason the connection may be misleading or coercive or even dangerous for that individual—as articulated in films such as *Enemy of the State* (1998), *The Matrix* (1999) and *Minority Report* (2002).

So the connection, the connectedness, which is signified by touch may be useful and positive in enabling us to relate to each other, other beings and objects, and to reflexively position ourselves in the world, but where it occurs without full knowledge of the individual subject it may be harmful and disabling. This is even more critical when we consider the related meaning or significance of touch, which is engagement.

9.4 Engagement

This meaning of touch encompasses the polysemy of touch as a physical, emotional and intellectual practice. Engagement, derived etymologically from the notion of a pledge or binding, captures the sense of touch as 'being with': touching as a way of contacting—being with—physically; as a way of feeling, empathizing—being with—emotionally; as a way of understanding, knowing—being with—intellectually. This meaning of touch places the toucher in an intimate relation to the touched, an acceptance of 'being with' that creates the opportunity for an empathetic relationship between the two.

Touch between individuals signifies engagement when it is accompanied by other practices—visual, verbal, aural, kinesthetic—that locate the contact as intentional; unlike, for example, the contact on a crowded train that signifies connection only. Again, the related practices are culturally and socially located. So, for example, a western man engaging with a close male friend may shake hands, embrace, put his arm around his friend's shoulders—again depending on factors such as class and age—but he is extremely unlikely to walk along holding his male friend's hand. To do so in western cultures would signify that their relationship is an intimate one; that they are lovers as well as friends. On the other hand, in other cultures hand-holding between male friends does not have this meaning; it signifies friendship. Further, the touch itself is further located by cues such as conversation, movement in relation to the other person, smiling and other visual responses. In this way, participants are able to specify to their own satisfaction—and according to the tactile regimes of their society and/or culture—the engagement signified by the touch.

Sometimes, however, this engagement may be problematic. Consider, for example, this description by a U.S. Army spokesman of their prototype Land Warrior uniform, a technologically-enhanced battle suit: "Armed with this technology the soldier becomes 'a totally, 100 percent integrated system… The computer… basically control[s] and manage[s] all the subsystems he's wearing'… his body is transformed into a personal-area network, and becomes a node within the larger network" [19]. This description raises serious questions about the nature of the engagement created by the soldier's wearing of this uniform and its successors. It also accords with the British Army's description of the soldier as a "biomechanical platform" [20]. In both cases the individual wearing a technologically-enhanced uniform is engaged in an assemblage over which control is distributed. He (or she) has access to many applications that provide access to information and communication, as well as others that protect him/her from attack; however, others also have access to the uniform—medical staff and command staff. This begs the question of what level of control is available to the individual wearing the suit and how his/her engagement should be configured. Furthermore, the bodily touch of these technologies creates a different kind of engagement, less driven by notions of individual freedom and control and more accepting of the role of the embodied subject as a nodal point in a network—social, economic, political, cultural. The 'being with' or engagement enabled by the touch (of the uniform) incorporates the wearer into a network that is outside her/his control, yet as noted earlier in the discussion of engagement, the nature of engagement is dependent on context. If the wearer cannot determine the context, then the engagement may have negative consequences (physical, emotional, intellectual) for the individual.

By contrast we might consider Cute Circuit's Hug Shirt—the shirt that incorporates electronics that enable parts of the garment to contract and warm up, so that the wearer feels they are being hugged. As the Cute Circuit web site explains: "Sending hugs is as easy as sending an SMS and you will be able to send hugs while you are on the move, in the same way and to the same places you are able to make phone calls (Rome to Tokyo, New York to Paris)" [21]. The immediate context for the engagement, as described by the designers, includes permission from the receiver for the hug, so the engagement signified by the touch (of the Hug Shirt) is under the control of the shirt's wearer. For this reason Cute Circuit believe that users can engage productively with the Hug Shirt, using it to reinforce relationships when the owners/wearers are separated geographically because of work or illness. Negative responses to the Hug Shirt and similar applications concern the possible replacement of human touch—and its associated meanings—with the cold (which is to say, semiotically empty) touch of the machine. Francesca Rosella of CuteCircuit approaches this concern quite differently, locating the context of the engagement—the human re-choreographing of space/time that is enabled by mobile technology, allied with evoked memories of past hugs between the participants. In this context, she argues, the human users retain control of the exchange and use the technology to enhance the human engagement signified by touch. This argument rests on an informed awareness of the technology and its capabilities, which is related to another significance of touch—contiguity.

9.5 Contiguity

Touch signifies contiguity when we become aware of the boundary that separates us from others, objects, and the world around us. I am going to discuss this meaning of touch as contiguity by reference to new technologies where the notion of boundary—and particularly of boundary-crossing—creates considerable unease.

There are two, very different approaches to the design of the user interface: one supports the minimizing, even disappearance, of (the perception of) the interface [22] the other argues for the foregrounding of the interface and its materiality [23, 24]. For the former, success is measured by users' incorporation of the interface into their everyday lives, so they no longer consciously register it every time they use the technology. In a sense they no longer register the boundary between themselves and the technology and so lose the sense of contiguity. For the latter group, this loss of contiguity is potentially disempowering for the user; without a sense of boundary, they argue, users do not understand that their actions, behaviours and responses are shaped by the parameters or boundaries of the technology they are using but instead naturalize the interaction as a product of their own actions.

This accords with the notion of 'seamfulness' in technology design, derived from the work of Mark Weiser who saw the awareness of 'seams'—of the suturing of the technology into everyday life—as enabling consumers to adapt the devices to local conditions [25–28]. It is only when we try using a wireless interface in a black spot, for example, that we are reminded that we are using an interface—when the seams show. We might modify this notion and suggest that the recognition of contiguity prompted by a seamful interface is also 'seme-ful' in that it draws the attention of users to the interface and hence to the ways in which it 'makes meanings'. So, for example, we might explore the ways in which the software we use everyday has inbuilt assumptions and protocols that delimit the ways in which we use it and so the meanings we can make.

For some designers this seamfulness is extremely important in that it also reminds users not to ignore the seme-ful possibilities of the interface itself. Franzisca Schroeder and Pedro Rebelo write eloquently about the importance of maintaining this awareness of the interface by reference to the interaction of musician and instrument: "This performer/instrument environment should be conceived as one of participation and engagement, in which the instruments themselves suggest to us specific ideas of their textures and materiality. This means that the performer only becomes acquainted with the "thing" at hand by being able to test boundaries, negotiate subtleties and uncover threshold conditions" [29]. The 'threshold conditions' are the delimiting factors that govern the nature of the interaction enabled by the interface and they are accessed by touch—the performer's touch on the instrument, the user's touch on the keyboard, the (wearable) technology touching the body of the user. This polysemic 'touch' enables recognition of the boundary between self and other, that is both fundamental to the knowing of self and enables appreciation of the other.

When touch is interpreted as contiguity, we become aware of the boundary between self and other and so able to locate the specificity of the other. As Schroeder and Rebelo note, without this awareness we would not have "Jimi Hendrix's guitar feedback, John Coltrane's unique saxophone sound or John Cage's prepared piano" [29]; we might have technically perfect music but it would have no soul. The delimitations of the boundary are crucial to our creative engagement. In the same way we might argue that the awareness of the boundary between ourselves and others enables the rich, delicate and creative exploration of the possible relationships between us. Further, as the supporters of seamfulness (seme-fulness) argue, this awareness of boundaries also alerts us to the conditions under which the connection takes place: that this connection is not totally under our control, but is determined by the conditions that establish the interface—whether this is the materiality of technological devices, the software programs we use, or the cultural assumptions of individual subjects.

9.6 Differentiation

This is closely related to another major signification of touch—differentiation; touch signifies the difference between the self and the other, beyond the boundary. We deploy touch in this way throughout our lives, to learn about other people and other objects [30–32], locating them within a network of experiences and meanings that enables us to use them effectively and to understand them. If we recognize the specificity of the other, then we are able to create an interaction that utilizes the full potential of that other.

Touch both connects us to technology and differentiates us from it, locating its specificity as different from our own being. As Schroeder and Rebelo write about the potential of textiles to act within a technological assemblage: "Fabrics offer a culturally rich platform for technological intervention. Let us start from the complex relationship between touch and textiles, and resist the reduction of such potentially engaging research to functional and instrumental applications. Let us think of what type of musical instrument a silk scarf might become, before incorporating a keyboard into it" [29]. If we ignore the specificity of the other—person, object, material—then we will not perceive the potential it brings to an encounter and the meanings it offers. At the same time, we will not understand ourselves, as we will not have the nuanced understanding of difference to help us locate our own specificity.

Much has been written about the way that Ron Mueck's hyperreal sculptures evoke the sense of touch [33]. The visual contradiction that characterizes Mueck's work is the disparity between the sculptures' hyperreal surface and their size; the works are usually either very big or very small. For reasons of safety and conservation, Mueck's relatively fragile works cannot be touched by viewers and yet people crowd around the works and frequently reach out to touch or pat them. The most likely purpose of this touch is to differentiate—human from sculpture, organic from non-organic, real from non-real. Whether or not the viewer's desire is physically realized, this (impulse to) touch engages the viewer with the materiality of the sculpture, identifying how this work has been created, as well as how and why that image is not real, not-us.

When applied to our engagement with technology, we find the same potential for deconstruction. As Schroeder and Rebelo note, we can interact reductively with technology to produce purely instrumental or functional applications. Or we can understand technology by discerning its specificity, differentiating ourselves from it: "Technology is a mode of revealing. Technology comes to presence … in the realm where revealing and unconcealment take place, where *aletheia,* truth, happens." [34] The touch that differentiates is not a rejection or an othering, but a way of recognizing the truth of the other in its specificity and distinctiveness—and in the same moment, of recognizing our own specificity and locatedness, our being-in the-world.

9.7 Positioning

Physically, touch creates an awareness of our location in space/time through embodied engagement with the world around us. The proprioceptive and vestibular senses are internal touch senses that enable us to position our bodies in space, even without visual stimuli, and to achieve equilibrium or balance.

Consider how this operates in relation to Ron Mueck's work—for example, his large sculpture, *Wild Man* (2005), a seated figure that is over 9 ft high (if the figure stood, he would be about 15 ft high). Traditionally, large sculptures—monumental works—have been equated with power. Sculpture occupies the same space/time continuum as its viewers, challenging them to compare their own occupation of space/time with that of the work. With a large work the viewer can only see the whole work by looking up at it, a posture associated with the acknowledgement of power. In this way the physical positioning of the viewer in relation to the work articulates the meaning of the work. The Wild Man of Mueck's work, however, does not look powerful; he looks terrified, gripping the sides of his chair in a paroxysm of fear. This Wild Man is not a hero, but a being who is 'other', rejected by society like the traditional Wild Man of European cultures or a contemporary vagrant—and yet Mueck's work argues his importance to our understanding of ourselves. By contrasting the appearance of the figure with the traditional reading practice of a large sculptural work, Mueck challenges us to consider how and why we make judgments about the worth and value of the people we encounter. In Mueck's deconstructive work it is our treatment of the 'othered' rather than our praise of the successful that will stand as a monument to our society. This reading of the work demonstrates how touch signifies positioning in multiple senses—not only physically, but as an embodied practice involving also emotional, intellectual and spiritual meanings.

This understanding of positionality is crucial for the development of a range of technologies, including wearables and locative media. For example, how is the soldier within the U.S. uniforms positioned by that technology? Is she/he already located in such a way that any rejection of directives that may be delivered to and via the technology is impossible? Can the individual inside the uniform reject an unethical order or is this technology progressively negating the distance between the individual's own moral and ethical positioning and that of those who control the technology? In our everyday lives we use various forms of locative media to find shops and restaurants and to locate our friends. To what extent does use of locative media applications already implicate us in a consumer-driven ethos that contradicts many of our stated values? And how easily do locative applications become a way of secretly tracking people? Again, is our positioning—moral, ethical—undermined by the seductive touch of this interface? The alternative is that we may use touch deconstructively to understand how we are positioned by the technologies we encounter, which reflexively makes us more aware of our own social and cultural, moral and ethical positioning.

9.8 Semefulness

Touch is seme-ful in that it is full of meanings (physical, emotional, intellectual, spiritual), which are socially and culturally specific and located. Far from being a simple, muscular response or action, touch locates us in the world, connects us to each other, and enables us to operate effectively as embodied individuals and as social subjects. By mapping some of the key meanings generated by touch—connection, engagement, contiguity, differentiation, positioning—this paper identifies the possibilities and problems created by new touch-based technologies—both those that deploy the user's hand (digital) touch and those that touch the body of the user. Conceptualizing touch as a semiotic practice enables us to analyze our interactions—with each other and with objects (including technologies)—and so become self-aware, reflexive individuals, able to trace the determinants and consequences of our actions, for ourselves and for others.

The first step: unpick the semes.

References

1. Touch Lab website: http://touchlab.mit.edu/oldresearch/index.html
2. Manning E (2007) Politics of touch: sense, movement, sovereignty. University of Minnesota Press, Minneapolis
3. Classen C (1993) Worlds of sense: exploring the senses in history and across cultures. Routledge, London
4. Howes D (ed) (1991) The varieties of sensory experience: a sourcebook in the anthropoolgy of the senses. University of Toronto Press, Toronto
5. Howes D (ed) (2005) Empire of the senses: the sensual culture reader. Berg, Oxford
6. Howes D (2003) Sensual relations: engaging the senses in culture and social theory. University of Michigan Press, Ann Arbor
7. Pink S (2009) Doing sensory ethnography. Sage, London
8. Stewart K (1996) A space on the side of the road: cultural poetics in an "other" America. Princeton University Press, Princeton
9. Stewart K (2007) Ordinary affects. Duke University Press, Durham
10. Classen C (ed) (2005) The book of touch. Berg, Oxford
11. Paterson M (2007) The senses of touch: haptics, affects, and technologies. Berg, Oxford
12. Classen C (2005) Touch in the museum. In: Classen C (ed) The book of touch. Berg, Oxford
13. Courtauld Institute, London. 2008. Sculpture and touch, viewed at http://www.courtauld.ac.uk/researchforum/archive/2008/summer_2008.shtml
14. Serres M (2008) The five senses: a philosophy of mingled bodies (I), (trans: Sankey M, Cowley P). Continuum, London
15. Cranny-Francis A (2008) Touching skin: embodiment and the senses in the work of Ron Mueck. In: Anderson N, Schlunke K (eds) Cultural theory in everyday practice. Oxford University Press, Oxford
16. Pease A, Pease B (2006) The definitive book of body language. Bantam Dell, NY
17. Reiman T (2007) The power of body language: how to succeed in every business and social encounter. Pocket Books, NY
18. MacGregor N (2010) A history of the world in 100 objects. Allen Lane, London

19. Viseu A (2003) Simulation and augmentation: issues of wearable computers. Ethics Inf Technol 5(1):17–26
20. Cranny-Francis A (2007) Ecce techno, or, suiting the biomechanical platform: immersion and contemporary embodiment. Visual Commun 6(2):156–169
21. Cute Circuit web site: http://www.cutecircuit.com/products/thehugshirt/
22. Ishii H, Ulmer B (1997) Tangible bits: towards seamless interfaces between people, bits and atoms. In: Pemberton S (ed) Human factors in computing systems: CHI 97 conference proceedings. ACM, NY; Addison-Wesley, Harlow
23. Kettley S (2008) Peacocks and wallflowers: (in)visibility with digital jewellery. Visual Commun 7(3):303–316
24. Schroeder F, Rebelo P (2007) Wearable music in engaging technologies. AI Soc 22:85–91
25. Andersson K (2007) Seamful design in a seamful society. SIDER 2007, the 3rd scandinavian student interaction design research conference, Ronneby, 8–9 Mar 2007
26. Chalmers M, MacColl I (n.d.) Seamful and seamless design in ubiquitous computing. Comput Sci, University of Glasgow, Glasgow.
27. Chalmers M, MacColl I, Bell M (2003) Seamful design: showing the seams in wearable computing. The lnstjtution Of Electrical Engineers, Stevenage
28. Barkhuus L, Polichar VE (2010) Empowerment through seamfulness: smart phones in everyday life. Pers Ubiquit Comput 15(6):629–639
29. Schroeder F, Rebelo P (2007) Wearable music in engaging technologies. AI Soc 22:85–91
30. Montagu A (1971) Touching: the human significance of the skin. Harper and Row, NY
31. Heller S (1997) The vital touch: how intimate contact with your baby leads to happier, healthier development. Henry Holt, NY
32. Paterson M (2007) The senses of touch: haptics, affects, and technologies. Berg, Oxford
33. Greeves S (2003) Ron Mueck–a redefinition of realism. In: Bastian H (ed) Ron Mueck. Hatje Cantz, Ostfildern-Ruit
34. Heidegger M (1977) The question concerning technology and other essays, (trans: Lovitt W). Harper and Row, NY

Chapter 10
Conceptualisation of an Intelligent Salutogenic Room Environment

Catherine Dalton and J. D. Harrison

Abstract Human functioning in the built environment is affected by the degree of "fit" or congruence between a user and her/his surroundings [1, 2]. By extension, the culture of care-giving and physical environment of care settings are inextricably linked. This conceptual model, developed in the context of the MyRoom project [3], is based on analysis of prerequisites for psychosocial congruence, drawn from theories of environmental psychology [4, 5], and from the evidence-base [6]. The model examines how these requirements may be addressed through architectural design enhanced by ubiquitous affective computing, integrated into the built environment, to maximise person-environment fit in healthcare settings. Where specific user needs, arising from cognitive and physical impairment associated with ageing and dementia, are not fully met by the built environment, these needs may be addressed through affective computing. This is to be achieved by means of real-time processing of data from an integrated system of Body Sensor Networks and Room Sensor Networks. This paper describes in detail an adaptive salutogenic single-user room in an elderly care setting, as a template of how an environment responsive to a user's physical and emotional state might be realised, promoting *salutogenesis* [7] through optimal congruence. Psychosocial congruence, on which this paper focuses, is enhanced through actuation of multisensory applications designed to provide appropriate stimulation. Recent research on affective computing for children with ASD may be translational [8, 9]. A majority of elderly persons in residential care have some form of dementia [10]. This implies that design of residential care and dementia care environments for elderly people is effectively inseparable. Architecture, further enhanced by ambient technologies, has the capacity to act as a major, and timely catalyst for a radical re-thinking of the culture and environment of care.

C. Dalton (✉) · J. D. Harrison
Cork Centre for Architectural Education, Cork, Ireland
e-mail: cathydalton2306@gmail.com

P. Breedon (ed.), *Smart Design*,
DOI: 10.1007/978-1-4471-2975-2_10,

Keywords Salutogenesis · Affective computing · Healthcare · Dementia · Responsive environments

10.1 Introduction

The context of the aging demographic in Western society, and the consequent requirement for provision of suitable environments, including residential care, for an increasing number of older people with dementia forms a backdrop to the research [10].

Human behaviour takes place in physical settings that influence behaviour and functioning in complex and sometimes unexpected ways. All buildings affect their users, whether intentionally or not. Environment affects not only physical and functional tasks, but also psychological and social processes, ranging from performance on cognitive tasks to emotional well-being [6]. Good design is conscious of its power to exert such influence, and uses it carefully. Stress, including stress arising from deficits in environmental fit, is recognised as a contributor to both psychological and physical illness [11], including Alzheimer's Disease (AD), where it has been found to be linked both to onset and progression. Research has indicated that stress is related to the onset of dementia-related neural changes, [12], and that cortisol has a particular significance [13]. This may explain the greater prevalence of AD in people with depression or increased levels of anxiety [14]. An environment designed to minimise and alleviate inappropriate stress is therefore an essential prerequisite for elderly care. This paper provides a detailed framework for design of a single, own-room setting congruent with the needs of an elderly user, in a residential setting. Theories of congruence [1], or environmental fit [5], have been summarised in a previous paper [15] setting out basic principles for design of an intelligent and responsive healthcare environment. A successful environment supports not only survival needs and instrumental tasks, but also psychosocial functioning [4].

Adaptive architecture has to date almost exclusively involved adaptation to environmental variables, frequently through kinetic structures. Other adaptations in architectural contexts, created in response to human presence or activity, are often conceived of solely as creative interventions or performance art. This model proposes an alternative adaptive architecture, enabled by embedded wireless sensoring networks, which through them becomes responsive, in real time, to an individual user's emotional and physical state. Direct kinetic response to physiological data has been tentatively researched in ExoBuilding [16]. In the context of architectural theory, the MyRoom proposal [3] represents not only an extension of theories of responsive architecture as described by Nicholas Negroponte [17], but a fusion of research in affective computing and responsive environments, creating a new paradigm for user-centred responsive architecture, generalizable to all building types. In the MyRoom model, physical characteristics such as lighting, temperature, colour and images are altered to provide a constantly variable environment best suited to an individual's needs at a given time. However, it is in care settings, where users are likely to have some degree of physical and/or cognitive impairment, where this approach stands to deliver the greatest benefits. The increasing proportion of elderly people, and the inevitable increase in incidence of dementia provides an impetus to conceive a new type of responsive architecture that supports and maximises independent functioning and quality of life for older people. The following recommendations on design of a salutogenic (health-promoting) [7] single-room environment are derived from Heerwagen's summary of attributes of buildings which are supportive of wellbeing needs [2, 18]. In the proposed model, integrated sensor networks are used to enhance those attributes.

The restorative capacity of natural environments is well–documented [19, 20]; how design can replicate this requires detailed consideration. *Connection to nature and natural processes* can be facilitated in a number of ways. The most obvious means in the design of a resident's room is through carefully considered design a placing of windows, to optimise views and natural light. Cill heights, relationships to external spaces and resulting available views, position and type of external planting, and internal room layout all should be given detailed consideration. The physical characteristics of the user must also be taken into account: an older person in care may have restricted mobility or be confined to bed or a seated position for long periods. Seasonal changes may negatively, as well as positively affect external views available from a single room. There is therefore a significant case for provision of natural forms or imagery within the room, to complement availability of external views of nature. This model proposes that these be provided as multimedia applications based on natural images, initiated in response to indications of stressed states, or designed to provide stimulus when appropriate.

Physical access to external spaces can also provide contact with nature, and is commonly provided from communal areas in care settings. The possibility of direct physical access to outdoors from an individual resident's room should be considered, and may be afforded if the external space is suitably configured and secure. This is still uncommon in residential care design, though single-storey buildings are the norm in many countries (Ireland, UK, US). Access might be provided via "protected" personal space, e.g. a small patio/paved area with seating. In a multi-storey building (common in continental Europe) a balcony or winter garden may be provided. The provision of such access also increases freedom of individual choice and supports a variety of situations. Again, a frail elderly user's physical status may mitigate against independent or spontaneous use of such access, particularly during inclement weather.

Provision of external views coupled with a variable internal environment also support needs for *sensory change and variability*. An individual room will of necessity have a fixed, often single orientation. This affects both quantity and quality (brightness, colour) of daylight entering, and can contribute to sensory variability. Not all individual rooms will enjoy an ideal orientation. In domestic design, in temperate latitudes, it is common to locate rooms so as to benefit from natural light at those times of day when the space is most used, e.g. living areas face south and south-west, while bedrooms face east. While communal spaces in residential care settings are often located to benefit from orientation, the same priority may not be applied to resident rooms, though they often provide the only available personal space for relaxation, privacy and self-actualization. The requirements for orienting a single space to suit all desirable activities can conflict. In the preferred domestic-scale care setting, orientation will be greatly influenced by a combination of siting and priority of function: in short, if communal spaces "get" the best aspect, bedrooms may be relegated to less favourable positions. Small-scale courtyard models, while secure, may prove restrictive in terms of orientation.

To support *behavioural choice and control*, the room must accommodate a variety of activities. This may require little more than adequate space provision. While Finnish residential care models provide residents with bed-sit type accommodation, as well as communal spaces, other models realistically allow only space for sleeping and grooming in the resident's own room; this is insufficient to meet needs for psycho-social congruence, as described. Though "household" type models are based on the scale and detail of domestic buildings, it should be borne in mind that the "household" is artificial: residents may have little choice about those with whom they cohabit. Where only shared spaces (e.g. dining and living-rooms) are available for socialisation, they do not provide the same levels of privacy as similar spaces in one's own home: it is not always possible to predict or control who else may be sharing that space. Visitors' rooms are often a sort of "no-man's land"; they can be impersonal, and not conducive to informal socialisation. Ability to socialise is influenced by availability of designed as well as unintended places to meet [21, 22]. Environment governs not only opportunities for socialisation but also types of socialisation facilitated: the level of privacy in a physical setting affects degree of disclosure in conversation, and the level of intimacy achievable. The resident's own room therefore takes on additional significance as a potential location for more intimate socialisation, for example with family, or for meeting carers or clinicians in a familiar and reassuring environment. By the same token, it may also afford space for privacy, relaxation and psychological restoration. If suitable space for such activities is not available, this will inevitably impact negatively on well-being. The scenario of bedroom doors left constantly open, on the basis that residents require constant observation, is commonplace. Sensoring allows the possibility of carers monitoring user status remotely, with processed sensor data streamed to a carer's pc, or made available on a "virtual" window immediately outside the room, without the necessity for a direct line of sight. The latter approach is also suited to acute care rooms, where it might replace an observation window, which places a patient continuously "on view". The fundamental prerequisite to support a variety of activities is sufficient space allocation: it is not possible to provide for all these needs while limited by areas defined by many contemporary care standards. On analysis, they are insufficient to support wellbeing needs, as they inevitably limit the environmental support achievable. Lack of space will inevitably result in activities being supported only in a limited or predetermined form, no matter how well-designed the context.

The possibility of multiple room uses can be supported by a variable internal environment, including artificial lighting designed for both functional and psychosocial needs. "Cycled lighting" can support circadian rhythms [23]; blue–green light therapy is effective for both seasonal and non-seasonal depression [24], and light may also be used to minimise "sundowning" symptoms [25]. Lighting systems should have the ability to be automatically activated, to compensate for user impairment. Given the particular role of stress in relation to AD, activation of environmental changes which can assist in alleviating stressed states have particular potential for increasing wellbeing. A visual or multimedia application specifically designed to reduce stress, which is activated in response to biosignals indicative of an inappropriately stressed state in an individual user, will also have the effect of externalising a user's emotional state, and may thus also potentially promote self-awareness, as well as de-stressing, not only by exposure to the imagery but also as a consequence of biofeedback [26], and clearly communicating that internal emotional state to an observer or carer. This is particularly valuable in the case of a resident who has lost the ability to verbalise, as is characteristic in AD.

Opportunity for regular exercise is catered for by provision of secure, easily-accessible external spaces; prompts for use might be incorporated in the room environment. Social support and sense of community are largely catered for outside the scope of an individual's room, though the room can provide space for some types of socialisation, as described above, and should facilitate personalisation, including use of a resident's own furniture, and choice of decoration.

10.2 Principles for Interventions in Environments for Elderly People with Dementia

Corcoran and Gitlin [27] developed a series of principles for interventions in environments for elderly people with dementia (PWD), providing a framework based on manipulation of the environment to match environmental press [28] with user competence. These principles are applied here to application and interface design. Reduced competence renders a person with dementia more vulnerable to environmental influences, with ensuing increased stress.

Availability[1] refers to accessibility of objects in an environment. Clutter should be avoided, and an individual presented with familiar objects specific to the task in hand. Frail elderly people often create "control centres" in their living-space with useful objects close to hand, both visible and accessible. Reducing *complexity* offers accessible environmental information. This supports an argument for minimalism in interface design [29], and for a "Disappearing Computer" [30] approach, where applications, sensoring and associated networks are subsumed

[1] words in italics refer to principles identified in [23].

into a familiar physical context. *Flexibility* regarding use is not desirable: objects with one clear function are preferred; conversely, controls with multiple options should be avoided. Applications for easy access to personal data (music, photos, and video) will have *Symbolic Meaning,* and help maintain self-identity. *Complexity* is related to the required skill level and number of steps needed to complete a task. Using an interface or application should be simplified or broken down into smaller steps to enable successful completion, and also designed so that only repetitive gross-motor activity is required [31, 32]. This suggests possible solutions to interface design, including simple gestural interfaces or large-scale, simplified touch-screen icons. Simplified, concrete verbal instruction might also be utilised. Supporting temporal orientation and reinforcing routine can facilitate user competence [32, 33]. In this context, provision of cycled lighting to reinforce circadian rhythms helps establish *temporal boundaries.* Seasonality might be reflected in lighting and by varying the colour palette of visual applications. It has been suggested that as stressful situations accumulate as the day progresses, a person with dementia experiences a lowered stress threshold, which can lead to an increased likelihood of a "catastrophic reaction", [34]: aggressive outbursts, or restlessness associated with "sundowning". This might be countered by providing environmental cues for relaxation and restoration, including actuation of stress-relieving applications at intervals based on the user's own behaviour patterns. PWD commonly make mistakes, but may not recognise them, and may be upset by their consequences. Flexibility enabled by responsivity and automation can permit a user to safely explore how an environment operates, without rigid "rules." The possibility of making a "mistake" with negative consequences can be minimised through thoughtful design intervention: for example, taps could automatically switch off when water levels in a basin reach a specific level. Prompts for activities of daily living should be incorporated. An identified need for low-stimulus activity ("daydreaming"), necessary for defining and maintaining an individual's identity [35] might be supported by responsive actuation of applications e.g. dimmed lighting, actuation of visual applications, personal digital data applications (photo albums, personal music collection).

10.3 Sensoring for Use in Care Environments for Elderly People

For longitudinal research on vulnerable subjects in live environments, current sensoring methods are often ethically unacceptable. EEG sensing, though it might provide invaluable data, is an example. The apparatus used, i.e. scalp electrodes, has changed little since its introduction over 60 years ago, and is unsuitable for use in long-term settings when, even in the limited context of a laboratory setting, it constitutes an additional stressor. In people with mid-stage dementia there is a significant likelihood of removal of any alien object, such as an identifiable wearable senor. The focus in development of wearable sensors for PWD must therefore focus on miniaturisation and unobtrusiveness. As most of the bulk of current sensors is made up of a battery-pack, an obvious research priority for wearables is development of smaller batteries, or of energy-harvesting technologies. While there has been considerable progress in terms of wearables, the term must be redefined in this context as a *non-contact sensor* which can be incorporated into the user's *own* clothing. This would seem to point towards sensors/energy sources which are both thin *and* flexible, or highly-miniaturised, which may require specific collaborative research in the area of materials science. Recent research points to an alternative approach of using only room sensor networks, in response to the stated preferences of elderly users: in short, sensoring integrated into the built environment [36, 37]. Kinematic and PIR sensor data, and video analysis of facial recognition and posture/movement can all be used to infer emotional state, as in recent research into applications for children with Autism Spectrum Disorders (ASD) [8, 9]. For ASD subjects, research is investigating applications which infer emotional state from heart-rate, anticipating aggressive outbursts, where there is no warning sign in terms of verbalisation or change of facial expression (a feature of ASD) [8]. It seems reasonable to infer that similar applications might be developed for PWD, to address inability to communicate, and response to stressed states, given the significance of stress in dementia.

10.4 Characteristics of a System Enhancing Congruence in a Care Setting

Advances in sensor technologies include increasing miniaturisation and development of non-contact bio-signal sensors. Coupled with development of more powerful and efficient wireless sensor networks, they support possibilities for intelligent adaptive ambient sensing networks. This model envisages responsivity through adaptation of an entire system over time to achieve optimal conditions for functional and psychosocial fit for an individual. As individuals respond differently to stress, any system designed to provide psychosocial support will need to "learn" what constitute optimal conditions for an individual, based on patterns of behaviour and observed responses to intervention. This approach also permits adaptation to progressive physical and cognitive decline, as system response to sensing will continually affect the underlying context/environment. Though complete autonomy is theoretically possible, direct input by users and carers may be incorporated to support needs for control. These inputs are of particular value in a set-up situation, until the basis for optimal conditions is established by machine-learning. System architecture will require integration of sensing data from which emotional affect can be inferred, including data modeling of complex behaviour and daily activities, and of the underlying physical context. Applications may be conceived of as prosthetic, designed to address compensation for specific cognitive and physical impairment. Constant and useful adaptation in response to data indicative of

Table 10.1 Summary of proposed ICT system characteristics for enhancing environmental fit

Ubiquitous	Embedded in room fabric
Affective	Capable for inferring emotional state
Minimalist	To minimise cognitive load on user
Coherent	Collaborative sensing, integrating bio-sensing data with contextual awareness
Intelligent	Capable for flexible reasoning to support complex ongoing decision-making, cognisant of component interactions
Self-adapting	Continually learning from occupant and context, to enable "best fit" for an individual, adjusting to changing user needs

changes in the occupant's psychophysiological status and/or in ambient conditions, implies system intelligence, which will also need to take cognisance of interaction between network components. The model based on this template will be tested iteratively in a prototype currently being realised through collaboration with NEMBES[2] partners. Further research on applications and interfaces will focus on detailed examination of user interaction (Table 10.1).

References

1. Kahana E (1982) A congruence model of person-environment interaction. In: Lawton MP, Windley PG, Byerts TO (eds) Aging and the environment: theoretical approaches, Springer, New York, pp 97–121
2. Heerwagen JH, Heubach JG, Montgomery J, Weimer WC (1995) Environmental design, work, and well being: managing occupational stress through changes in the workplace environment. AAOHN J 43(9):458–468
3. Dalton C, Harrison JD (2010) MyRoom: A model for an intelligent salutogenic room prototype for an elderly user MyRoom: a model for an intelligent salutogenic room prototype for an elderly user. Gerontechnology 9(2):206
4. Boyden S (1971) Biological determinants of optimal health. In: Vorster DJM (ed) The human biology of environmental change, conference proceedings. Blantyre, Malawi
5. Kaplan S (1983) A model of person-environment compatibility. Environ Behav 15(3):311
6. Dilani A (2008) Psychosocially supportive design: a salutogenic approch to the design of the physical environment. World Health Design, July 2008
7. Antonovsky A (1987) The salutogenic perspective: toward a new view of health and illness. Advances 4(1):S47–S55
8. Picard RW (2009) Future affective technology for autism and emotion communication. Philos Trans R Soc B: Biol Sci 364(1535):3575–3584
9. Poh MZ, McDuff DJ, Picard RW (2011) Advancements in noncontact, multiparameter physiological measurements using a webcam. IEEE Trans Biomed Eng 58(1):7–11
10. Torrington J (1996) Care homes for older people. Taylor & Francis, London
11. Antonovsky A (1996) The salutogenic model as a theory to guide health promotion. Health Promot Int 11:1

[2] NEMBES is an inter-institutional multi-disciplinary research programme investigating a "whole system" approach to the design of networked embedded systems. www.nembes.org.

12. Osborne F, Almeida X et al (2011) Stress acts cumulatively to precipitate Alzheimer's disease-like tau pathology and cognitive deficits. J Neurosci 31(21):7840–7847
13. Davis KL, Davis BM, Greenwald BS, Mohs RC, Mathé AA, Johns CA, Horvath TB (1986) Cortisol and Alzheimer's disease, I: basal studies. Am J Psychiatry 143(3):300–305
14. Wilson R, Evans D, Bienias J, Mendes de Leon C, Schneider J, Bennett D (2003) Proneness to psychological distress is associated with risk of Alzheimerís disease. Neurology 61(11):1479
15. Dalton C, Harrison JD (2011) Designing an adaptive salutogenic care environment. In: Proceedings 1st international conference on adaptive architecture, London, forthcoming
16. Schnadelbach H (2011) Physiolocial data in adaptive architecture. In: Proceedings of the 1st international conference on adaptive architecture, London, forthcoming
17. Negroponte N (1975) Responsive Architecture. In: Soft Architecture Machines, MIT Press, Cambridge
18. Heerwagen J (2006) Investing in people: the social benefits of sustainable design. In: Proceedings of the rethinking sustainable construction' 06, Sarasota, FLA, US
19. Ulrich RS (1979) Visual landscapes and psychological well-being. Landscape Res 4(1):17–23
20. Ulrich RS (1984) View through a window may influence recovery from surgery. Science 224(4647):420
21. Archea J (1977) The place of architectural factors in behavioral theories of privacy. J Soc Issues 33(3):116–137
22. Sundstrom E, De Meuse KP, Futrell D (1990) Work teams: applications and effectiveness. Am Psychol 45(2):120
23. Rea MS, Figueiro MG, Bullough JD (2002) Circadian photobiology: an emerging framework for lighting practice and research. Lighting Res Technol 34(3):177
24. Lee T, Chan C, Paterson J, Janzen H, Blashko C (1997) Spectral properties of phototherapy for seasonal affective disorder: a meta analvsis. Acta Psychiatr Scand 96(2):117–121
25. Mishima K, Hishikawa Y, Okawa M (2001) Randomized, dim light controlled, crossover test of morning bright light therapy for rest-activity rhythm disorders in patients with vascular dementia and dementia of Alzheimer's type. Chronobiol Int 15(6):647–654 (1998)
26. Shi X, Guo Y, Nakagawa Y, Morita M, Uchiyama H, Fukomoto I (2001) A study of dementia rehabilitation BF system by light-stimulation. Jpn J Biofeedback Res 28:79–87
27. Corcoran M, Gitlin LN (1991) Environmental influences on behavior of the elderly with dementia. Phys Occup Ther Geriatrics 9(3–4):5–22
28. Lawton MP (1982) Competence, environmental press, and the adaptation of older people. In: Lawton MP, Windley PG, Byerts TO (eds) Aging and the environment: theoretical approaches, Springer, New York, pp 33–59
29. Wren CR, Reynolds CJ (2004) Minimalism in ubiquitous interface design. Pers Ubiquit Comput 8(5):370–373
30. Streitz N, Nixon P (2005) The disappearing computer. Commun ACM 48(3):33–35
31. Levy LL (1987) Psychosocial intervention and dementia: I. State of the art, future directions. Occup Ther Mental Health 7(1):69–107
32. Berman S, Rappaport MB (1985) Social work and Alzheimer's disease. Soc Work Health Care 10(2):53–70
33. Oakley F (1988) Clinical application of the model of human occupation in dementia of the Alzheimer's type. Occup Ther Ment Health 7(4):37–50
34. Hall G, Buckwalter K (1987) Progressively lowered stress threshold: a conceptual model for care of adults with Alzheimer's disease. Arch Psychiatr Nurs 1(6):399
35. Rubinstein RL (1989) The home environments of older people: a description of the psychosocial processes linking person to place. J Gerontol 44:S45
36. Doyle J, Bailey C, Dromey B (2009) Experiences of in-home evaluation of independent living technologies for older adults
37. Bakkes S, Rijken M, Kr̈ose B (2011) Telemonitoring for independently living elderly: inventory of needs and requirements. In: Proceedings of pervasive health 2011. Forthcoming

Chapter 11
Development and Thermographic Analysis of Custom Seats for Wheelchairs

Rafael Cavalli Prestes, Elisa Marangon Beretta, Fabio Pinto da Silva, Wilson Kindlein Jr. and Vilson João Batista

Abstract The design of Assistive Technology products has different technological routes such as massive products or custom products. These products aim at optimize the altered bodily functions. With that it can re-conduct these people to varied social activities. In this sense, the present article focuses on contribute in the construction of new technological routes for the manufacturing of wheelchair custom seats. The methodology includes the application of concepts such as tri-dimensional digitizing, CNC machining and thermography. Those techniques have the purpose of generating and analyzing seats made from user's anthropometric data.

Keywords Product design · Assistive technology · Custom seats

---

R. C. Prestes (✉) · E. M. Beretta · F. P. da Silva · W. Kindlein Jr. · V. J. Batista
Universidade Federal do Rio Grande do Sul, Porto Alegre,
90035-190, Brazil
e-mail: rafael.81@hotmail.com

R. C. Prestes · E. M. Beretta · W. Kindlein Jr. · V. J. Batista
PgDesign, Porto Alegre, Brazil

F. P. da Silva
Departamento de Design e Expressão Gráfica, Porto Alegre, Brazil

W. Kindlein Jr. · V. J. Batista
Departamento de Materiais, Universidade Federal do Rio Grande do Sul,
Porto Alegre, Brazil

P. Breedon (ed.), *Smart Design*,
DOI: 10.1007/978-1-4471-2975-2_11,

11.1 Introduction

Nowadays there is a major deficiency in the development of products aim at people with disabilities. According to the United Nations, approximately 10% of the global population has some sort of disability [1]. This means 650 million people with disabilities.

The product design for Assistive Technology follows different technological routes with massive products or customized products. Assistive Technology is the term used to express a great variety of resources and services aimed at people with disabilities. These individuals need specialized services from different areas to facilitate their inclusion into varied social activities. An interdisciplinary team should provide those services. This team would have in its formation professionals involved with rehabilitation processes (such as doctors, physiotherapists, occupational therapists, psychologists, product designers and engineers).

According to Moraes [2] besides Assistive Technology services, other important area is the products and systems that substitute structures and body functions lost or absent. It can be defined in different areas. One of them is postural adequacy, where wheelchair systems can be placed. There are in numerous innovations in this equipment. The innovation that stands out most is the evolution of seats cushions and the adaptations of the dispositive for seated posture [3]. In general terms, the wheelchair is made in a standard size witch difficult the adaptation for users with peculiar postures. Although more sophisticated wheelchairs equipped with special features were created, they do not allow the adjustment to individual deformities.

Watson [3] asserts that there are few scientific studies that affirms which seat model or system is preferable. Until this moment little is known about which dispositive can extract maximum benefits to its user. In this sense, this study focuses on the development of technological routes to manufacture custom seats that satisfy the specific demands of each user. The paper was divided in materials, methods, results, and conclusions.

11.2 Materials and Methods

A case study was realized for the proposal of a seat dispositive development that increases postural adequacy. It was initialized by the evaluation of a 47 year-old male user with medical diagnosis of spinal traumatism, physiotherapeutic diagnosis of paraplegia. He presented diminish sensibility in the upper members and lateral thighs and absence of sensibility below these members.

Tridimensional data about the wanted shape were obtained through tridimensional scanning processes. In this case, the wanted shapes were the user's bodily parts such as hips and thighs. A tridimensional scanner Konica Minolta model

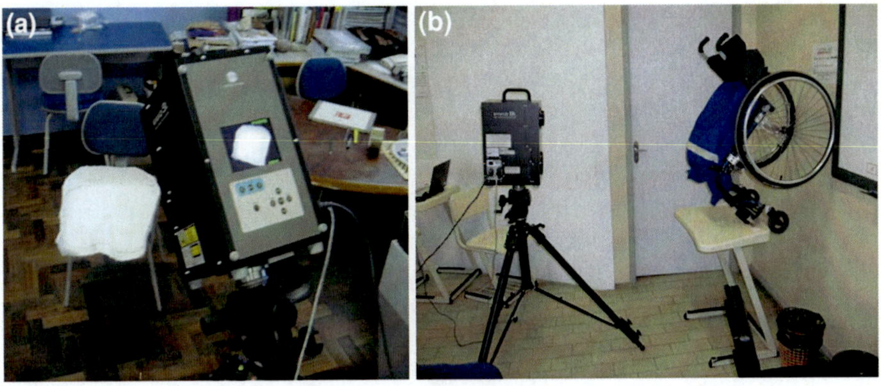

Fig. 11.1 Seat development. **a** Plaster cast scanning. **b** Vacuum cushion scanning

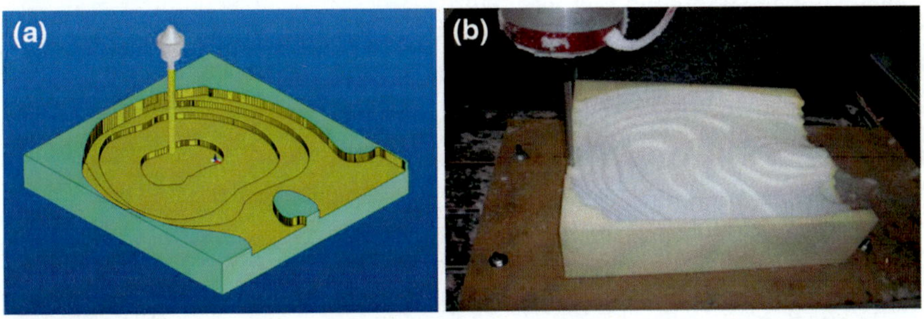

Fig. 11.2 **a** Milling strategy. **b** CNC milling

Vivid 9i was used for the data acquisition. The lenses used were wide lenses (focal distance 8 mm, resolution 0,096 mm, distance to object 1 m).

The anthropometric data of the first seat was acquired through a cast made of plastered bandage (Fig. 11.1a). The cast was moulded directly from the user's shape. The user was positioned seated in the wheelchair with hips and knees at 90° and ankles at 0°. The second seat was developed from anthropometric data obtained from a vacuum cushion Versaform, as in Fig. 11.1b, following the same position criteria.

The output files from the tridimensional scanning processes constitute itself of a point cloud (in witch can contain millions of points). These represent x, y and z coordinates that compose the scanned surface. After the 3D scanning, the data were processed using Raindrop Geomagic Studio software. For the processing were realized (1) assembly of the point clouds and (2) filter application. The filter application aims at reduce the noise inherent to the 3D scanning process. In addition, it can reduce the quantity of points in the cloud and maintain just the significant points to represent the object. Likewise, it can make the tridimensional model easily manipulable. After, the points from the 3D scanning were united in trios forming triangles. They generate a tridimensional mesh from the surface related to the user's thighs and hips. The portion of the data needed to manufacture the seat cushion is maintained and the others are excluded.

The archives were then exported to Edgecam software. In this software milling strategies were created for the seat manufacture. Two steps composed the milling strategy. One of them is the roughing (Fig. 11.2a) and the other is the finishing. The materials used were two blocks of flexible polyurethane foam with initial dimension of 380 × 360 × 100 mm and density of 45 kg/m^3. The process was made in the Tecnodrill Digimil 3D CNC equipment. The cutting tools used had a diameter of 10 mm ands were a double-flute mill with edge angle at 35°. As milling parameter it was used a 12000 rpm spindle speed, 2000 mm/min horizontal feed rate, 600 mm/min vertical feed rate and 10 mm cut increment (Fig. 11.2b). For the finishing was used 1200 rpm spindle rate, 1600 mm/min horizontal feed rate, 800 mm/min vertical rate and 15% stepover.

When the CNC was done and the seat was ready, it was chosen to quilt the cushion with a nylon fabric to waterproof the seat. This can avoid the users sweat to get in contact with the cushion. In order to analyze the thermal differences between wheelchair's standard seat and the custom seat manufactures it was used the thermography. This technique is used to verify through infrared radiation emission areas with major and minor heat concentration.

11.3 Results

The thermographic analysis of the seats was realized in a heated room at 23°C. It was analyzed three seats: the wheelchair standard seat, the custom seat made from the plaster cast, and the custom seat made from the vacuum cushion. The user was oriented to wear just underwear before executing the thermographic registers. He was positioned in a standard chair and stayed seated for approximately 15 min. He was then positioned in his wheelchair using the standard seat. Adjustments in his feet and arms were made to improve weight distribution over the seat. He stayed in this position for 20 min. When out of the chair the thermographic register was made. In this register two lines can be seen (lines E and F in Fig. 11.3). They extend themselves from the anterior part of the seat until the posterior part. They correspond to the tights distal and posterior areas until the hips. The Figs. 11.4 and 11.5 follow the same organization principle. Figure 11.3 corresponds to the seat manufactured from the data obtained through the plaster cast. Figure 11.5 corresponds to the seat manufactured from the vacuum cushion.

11.4 Discussions

The manufacture of custom seats according to the user's anatomy is an important tool in the prevention of tissues lesions. It can also help problems related to the spine biomechanics. A problem to generate custom seats is the attainment of the

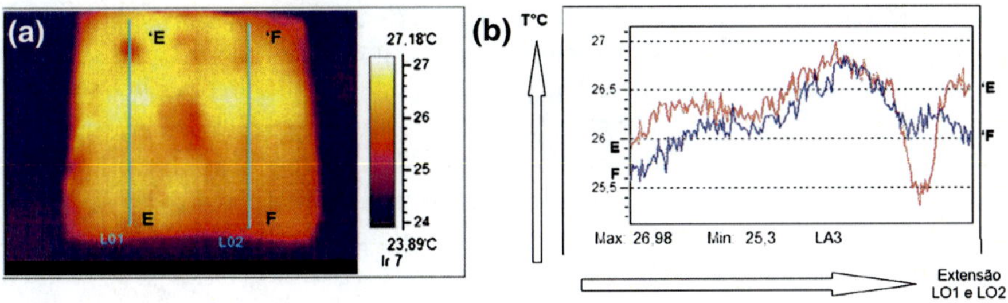

Fig. 11.3 **a** Thermographic register from the standard seat and **b** *Lines E* and *F*

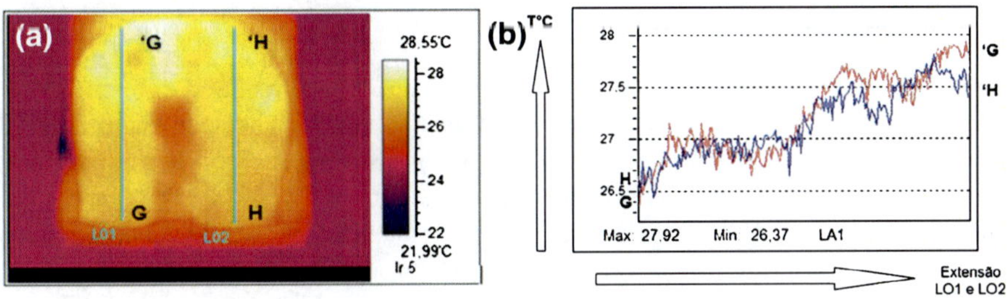

Fig. 11.4 **a** Thermographic register from the plaster cast seat and **b** *Lines G* and *H*

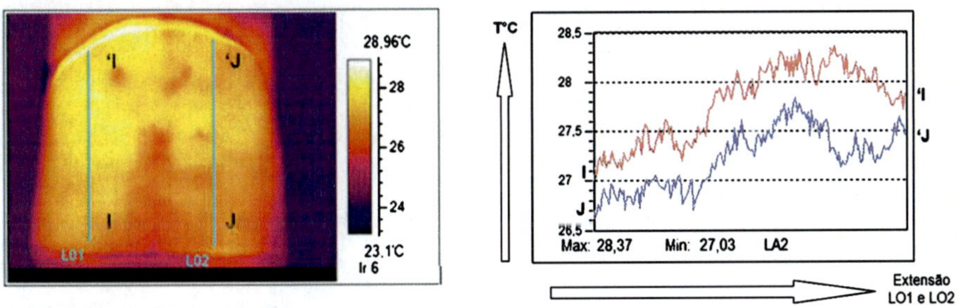

Fig. 11.5 **a** Thermographic register from the position cushion seat and **b** *Lines I* and *J*

bodily shapes and contours. This problem can be solved through the 3D scanning process. According to Silva et al. [4], this process allows the obtainment of 3D objects with major details such as curves and textures.

For Brienza et al. [5], the manufacture of a custom seat that reproduces the user's anatomy and made of a flexible material can contribute significantly to reduce the pressure scars. Before addition, it is supposed that a foam cushion with the user's contour can distribute evenly the applied forces made by the bodily weight through uniform strain during use.

The standard seat presents more thermal variation when analyzing the thermal differences traced by the thermographic registers. It can be seen that the standard seat showed a thermal variation of 2°C in the left side and a 1.5°C in the right side. The custom seat made from the vacuum cushion presents variation of 1.5°C through the two lines in the graphic. It should be noted that the differences maintained a pattern during its course. The seat manufacture from the plaster cast also had thermal differences of 1.5°C. These differences showed more linearity between sides. In this sense it can be affirmed that the seat with best thermal distribution was the one obtained from the plaster cast.

11.5 Conclusions

The technological route for this manufacture system is practical and efficient. After the results analysis the goal of this research was achieved. The seat manufactured from the plaster cast was the most effective as distributing the bodily temperature. It also can be used to prevent skin lesions.

It is believed that the developed equipment can improve the rehabilitation process. It can generate more comfort and life quality by acting in the prevention of cutaneous scars caused by thermal differences.

References

1. United Nations (2006) Some facts about persons with disabilities. United Nations, New York, pp 14–25
2. Moraes HS (2009) Projeto Conceitual de Sistemas de Assentos para Cadeira de Rodas: Uma Abordagem Sistematica. Dissertação de Mestrado. Programa de Pós Graduação em Design, Universidade Federal do Rio Grande do Sul
3. Watson N, Woods B (2005) The origins and early developments of special/adaptive wheelchair seating. Social Hist Med 18(3):459–474
4. Silva FP, Salvalaio CL, Kindlein JW (2010) Analise Comparativa de Processos de Digitalização 3D para Modelos de Partes Humanas. 9° Congresso Brasileiro de Pesquisa e Desenvolvimento em Design, São Paulo
5. Brienza DM, Geyer MJ, Karg P, Jan YK (2001) State of the science white paper on tissue integrity management. Wheelchair Seating, v

Chapter 12
Designing a Knit Methodology for Technical Textiles

Martha Glazzard and Philip Breedon

Abstract This project assumes a perspective developed from practice in textile making and design. To a textile practitioner, it becomes clear that in the field of technical textiles, often a radically different approach is used from that used in the traditional apparel industry. The knitted apparel industry has joint emphases on aesthetics and functionality, in both fabric and garment development stages (Eckert, Intelligent support for knitwear design, PhD thesis, 1997). In technical textiles, the emphasis often comes from an engineering point of view and is primarily concerned with function (Stead, The emotional wardrobe: a fashion perspective on the integration of technology and clothing, PhD thesis, 2005). These different design perspectives are capable of enormously different results, or can be unintentionally close to each other's disciplines. In weft-knitted textiles, the methods and several of the considerations used to make fabrics do not differ, whether the desired outcome is function-focused or aesthetically focused. In reality, it is always both of these things. The decisions faced in development of a garment are invaluable to any developmental textile work. They inform on shape, fit, quality and durability at every stage of the production process (Aldrich, Fabric, form and flat pattern cutting, 2nd ed, Blackwell, 2007). These considerations make differences in producing well-integrated technologies into textile forms and the difference between technology and/or functionality existing within a product rather than sitting on top as a separate entity. This paper talks of the early stages of a research project which attempts to delineate the approaches adopted when designing a technical textile, in order to take more account of tacit and intuitive knowledge which comes from textile as a design discipline. Already showing

M. Glazzard (✉) · P. Breedon
Nottingham Trent University, Nottingham, UK
e-mail: martha.glazzard2003@my.ntu.ac.uk

P. Breedon (ed.), *Smart Design*,
DOI: 10.1007/978-1-4471-2975-2_12, © Springer-Verlag London 2012

interesting results about discipline and methodology, the case study uses auxetic materials [those which expand in a transverse direction to that of the stretch (Lakes, Science 235:1038–1040, 1987)] as a case study. Design considerations utilise joint emphases on form and function-led methodology.

Keywords Methodologies · Knit design · Auxetics

12.1 Introduction

Knitted textile design is a complex procedure. Inputs range from 'gut instincts'—qualitative assessments, aesthetics and tactile considerations—to quantitative considerations such as trigonometry, fit, form and function. Knit design as a methodology is unusual in textile design as traditionally, it has always been possible to design the form and the function as well as the 2D and the 3D simultaneously [1]. Knitted outcomes require parallel thinking about the qualitative and the quantitative elements. Literature on knit design methodologies is mostly found in texts addressing textile design as a broader subject. This, however, is not of much compensation due to the comparative lack of theoretical framework present in textile design, leading to its common omission from research [2].

Increasingly, in the discipline of *textiles* there is a separation of engineering thinking from design thinking [3]. Let us define engineering thinking as being that concerned with quantifiable data: the dimensions, the mechanical properties and the performance properties. The design thinking covers the aesthetic, tactile and instinctive judgments on quality. Definitions of both engineering and design include reference to tacit knowledge acquired through professional and academic experience.

12.2 Methodological Contest

A major dispute between the design and engineering disciplines lies in the methodology. Quantitative and qualitative methodologies are rarely mixed: attempts to do so risk alienating and confusing potential audiences [4].

In knit design there is little apparent history of methodology as a formal concept. Taught as an industrial and creative practice under headings of 'knitted textile design' or 'fashion knitwear design', there can be many strands to the purpose of the designing and therefore to the methods undertaken.

Knit design can take on attributes of fashion design, fine art, product design and textile engineering, but there is no reason that the outcome achieved from any of these methodological standpoints could not result in a similar end product.

12.3 Defining a Project

This project implies a holistic approach to designing textiles that incorporates both qualitative and quantitative considerations. This does not seem unusual in the context, as it may be deemed to be the most logical way to approach any kind of design. When being 'creative', major decision-making can come down to intuition and tacit knowledge. By using a justification that if it doesn't look right, or feel right, then it probably isn't right, it is second nature to attribute sensual judgements to our overall sense of 'liking' something. These are easy to describe to others and ourselves. The uses of quantitative assessment are also obvious, there are always parameters—be they size, shape, cost, time—that are considered in the design or making process. The quantitative is in place to assess whether or not those criteria have been met.

If we look at knit design as a methodology, then there are several approaches to each project, some of which are summarised below:

1. Working towards a specific application—making drawings, samples, mock-ups and final product.
2. Working with fabric to inform application (draping or modelling)—making small and then large samples and working with the shape and drape of the fabric to inform the application.
3. Designing around patterns/textures/fabric shapes—using one as a starting point and allowing the design to complement that starting point.
4. Designing around yarns—allowing the yarn to dictate the fabric weight and structure. Information from drape and fabric qualities is then fed into the design process.
5. Designing around feature information or details, such as make up information, seams etc.
6. Designing to specification—cost, market, trend…

This means that the process of designing can shift and develop in many ways, depending on what the designer chooses or is required to give most emphasis to. All of the knit design process relies heavily on both quantitative and qualitative assessments throughout.

By using a case study to build parameters around the subject covered, it should be possible to see this mixed-method thinking in practice. Through the case study, the methodology used in knit design processes should become clear.

12.4 Research Design

The design of this research involves practical and theoretical work in parallel strands. The practical involves knitting auxetic (those which expand in a transverse direction to that of the stretch [5]) fabrics with the intention of emphasising the auxetic effect in the fabrics using regular knitting machinery and design considerations. The theoretical aspect will extract information about the methodological practice involved in knit design. This will be collected via reflection *in* action and reflection *on* action [6].

12.5 First Stage

Auxetic materials were chosen, among other reasons, for their suitability to be reproduced with knitted structures, as demonstrated by Liu et al. at the Hong Kong Polytechnic [7, 8]. Liu et al. have ongoing funding to explore auxetic weft-knitted textile structures (Fig. 12.1). Work they presented in November 2010 shows how a simple relief pattern structure can create auxetic behaviour in a woollen, knitted fabric. This demonstrates how traditional, knitted patterns and designs have shown worth in a scientific field. Typical to the field, Liu et al. present their work as an engineering paper with the relevant formulae and quantitative information. What was noticeably absent from Liu et al.'s study was any recognition of the aesthetic and tactile qualities of the fabric. The new research at Nottingham Trent University encourages visual development of different structures, where knit and purl stitches interplay. Knowledge from experience of fabrics, which would create 3D folding or gathering effects is also encouraged. Several samples have been tried, which are pattern variations on Liu et al.'s samples, patterns that have been conceived for their geometric pattern appeal or patterns that had been used for aesthetic design purposes in the past. One of the most successfully auxetic fabrics, showing promising auxetic effect in the y-axis with comparatively little extension in the x-axis is the 'basket-weave' (Fig. 12.2)—a chequer board pattern in rectangles of knit and purl—which is very common in hand and machine knitting.

The second alteration made to Liu et al's project was the addition of lycra alongside the woollen yarn to increase the retraction of the samples to the original pleated form. This is a tactic borrowed from clothing design to improve fit of knitted garments to the body [9]. The samples with the lycra had a tighter visual appearance and feel, as well as a faster and more convincing return than those with only wool.

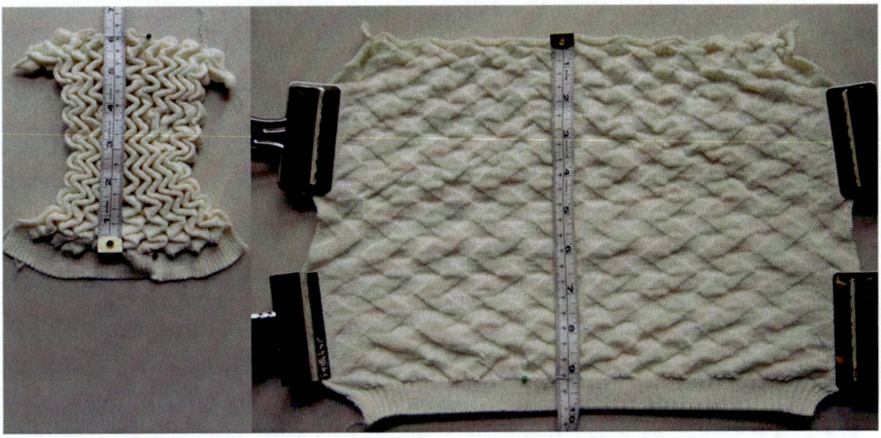

Fig. 12.1 Replication of Liu et al. sample showing stretch in the x-axis and expansion in the y-axis

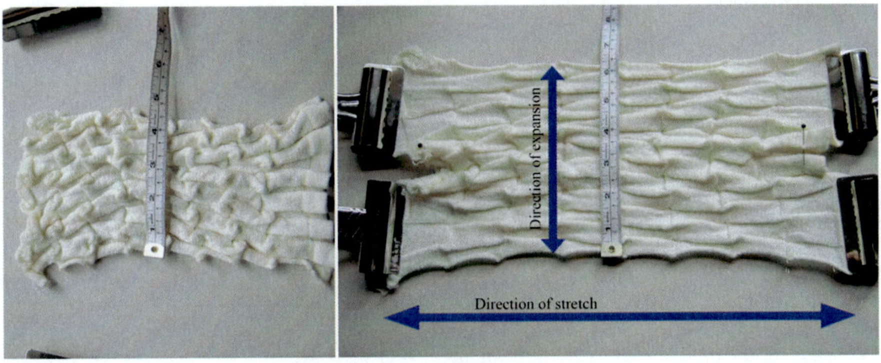

Fig. 12.2 'Basket-weave' sample showing auxetic effect

12.6 Conclusion

In conclusion, we have seen that knit design is worthy of classification as a methodological subject in its own right for the following reasons:

- As a textile, knit is unlike other textiles in the simultaneous aesthetic, functional and 2D, 3D considerations.
- The design is not necessarily pre-determined by any one factor, and a number of outcomes may be created.
- The methodology used in the design process is not always pre-determined by the end result.

The knitting process in this study uses conventional technology and common yarns to create auxetic effects. These auxetic behaviours are demonstrated to varying degrees and the fabrics vary greatly in appearance and visual properties.

The mixing of methodology has shown some potential of being problematic in so far as there is the possibility of discouraging interest and contribution from different disciplines. The mixed methodology is not seen to be a problem in this author's view. Having come from a background where there is no formal idea of methodology [10], it is not surprising to the author that methodological boundaries cannot at this stage be clearly defined.

In creating auxetic textiles, the work covered so far has demonstrated that experiential knit design tactics can be used to expand engineered auxetic effects. Here, it is the knowledge learned from dealing with knitted fabrics for clothing and aesthetic purposes that has determined the direction for the sampling process.

There are many variables in the production of these simple auxetic fabrics, meaning the potential for future research is sizeable. Further work will aim to emphasise auxetic effects using knitted stitch structures. Aesthetic and tactile contributions can be weighed against auxetic effect to allow for a range of fabrics, which can be used for their functionality along with their appearance and other sensual attributes.

This work could potentially open out the use of auxetic materials to designers and artists, but also hopes to encourage function-focused practitioners to consider some of the aesthetic and tactile design approaches.

References

1. Eckert C (1997) Intelligent support for knitwear design. PhD thesis, The Open University
2. Bye E (2010) A direction for clothing and textile design research. Clothing Text Res J 28:205
3. El-Mogahzy Y (2009) Engineering textiles: integrating the design and manufacture of textile products. CRC Taylor and Francis distributor, Boca Raton
4. McCabe B (2006) Practice-based textile design research as a method to explore knowledge transfer frameworks. Nottingham Trent University, Nottingham
5. Lakes R (1987) Foam structures with a negative poisson's ratio. Science 235:1038–1040
6. Schön DA (1991; 1983) The reflective practitioner: how professionals think in action. Ashgate, London
7. Liu Y et al (2010) Negative poisson's ratio weft-knitted fabrics. Text Res J 80(9):856
8. Hu H, Wang Z, Liu S (2011) Development of auxetic fabrics using flat knitting technology. Textile research journal. http://trj.sagepub.com/content/early/2011/05/11/0040517511404594. Accessed 21 June 2011
9. Aldrich W (2007) Fabric, form and flat pattern cutting, 2nd edn. Blackwell, Oxford
10. Stead LJ (2005) The emotional wardrobe: a fashion perspective on the integration of technology and clothing. PhD thesis, University of the Arts

Chapter 13
Soft Products: Smart Fabrics in Product Design

Polly Duplock

Abstract A definition of 'soft products' is proposed: products designed by a product designer which traditionally would be made from hard materials, but have a substantial component constructed in fabric. As such, soft products are distinct from furniture design and the design of accessories by fashion designers–and from 'wearables', a field that continues to attract more attention and development. New fabrics with increased performance or added functionality have allowed product designers to consider new uses for textiles. Smart fabrics (with embedded electronic capability) have enabled 'soft products' to become realities. The Logitech KeyCase soft keyboard enabled by Eleksen's ElekTex fabric is one of the most significant 'soft products' that has made it to market, where the fabric has dramatically changed the functionality of the object. Beyond the rolling facility though, the keyboard has all the usual functionality of a keyboard and is styled to have the aesthetic language of a 'gadget'. This keyboard can be starkly contrasted with the original range of products that were developed by IDEO for the launch of the ElekTex technology. This project, entitled 'Fabrications', was speculative with the ambition of promoting the possibilities of the technology rather than developing market-ready solutions. Why have smart fabrics not begun to form a bigger part of the product designer's palette of materials and to revolutionise the way we design products? In this paper, the author explores this question by reviewing the landscape of smart fabrics currently available, discussing their relevance to product design, reflecting on the Fabrications project, contrasting cultural values in product and fashion design and describing a pioneering module with undergraduate product design students.

P. Duplock (✉)
Duncan of Jordanstone College of Art and Design,
University of Dundee, Dundee, DD1 4HT, UK
e-mail: p.duplock@dundee.ac.uk

P. Breedon (ed.), *Smart Design*,
DOI: 10.1007/978-1-4471-2975-2_13,

Keywords Soft products · Smart fabrics · Product design

13.1 Defining Soft Products

There seems to be no clear existing definition for 'soft products' and I would like to use this paper to propose one: 'Soft products are objects designed by a product designer, but instead of being made with the usual palette of hard materials such as metal, plastic or wood they have a substantial component constructed in fabric.' These objects could include domestic appliances, consumer electronics and lighting for example. This definition does not include furniture, because it has its own specialist discipline of furniture design or bags/accessories, which are generally considered by fashion designers. It also excludes what have come to be known as 'wearables'–an area that has received far more attention in industry and academia and in which there are more recent exemplars, such as the Nike+ wearables.

13.2 Smart Fabrics

In a recent presentation at Smart Fabrics 2011 Adrian Wilson, an analyst, writer and editor specializing in technical textiles gave an overall broad definition of smart fabrics as being "any fibre-based structure that can react to stimuli" [1]. This definition includes many 'enhanced fabrics' such as garments specially treated to be wrinkle free, spill or stain resistant, anti-fading, with moisture management or with odour control properties etc. Fibres and fabrics with a degree of shape memory, nano-enhanced materials or those which react to temperature changes are other examples of stimuli-reacting products which come under this definition.

Wilson then went on to narrow his definition referring to smart fabrics as "An umbrella term for a range of technologies, which in many cases fuse clothing or worn accessories with portable electronic devices, or in a broader sense put some kind of electronic functionality into other materials". This paper will restrict itself to this narrower definition i.e. textiles which have some form of embedded electronic capability.

Smart fabrics can currently be broadly categorized into four areas. These are heat, light, touchpad and sensor textiles [2]. There have been some success stories in each of these categories and for the purpose of this paper I will briefly introduce one company working in each of these defined areas:

Heat- ITP GmbH are a German company specializing in textile innovation who currently produce heated knitwear where electrically conductive yarns are incorporated in a mix together with normal textile yarns.

Light- Lumalive is a business venture being developed by Philips. A conductive yarn circuit and integrated LEDs are applied to fabric in a grid like structure resulting in a digital wearable display system.

Touchpad- ElekTex by Eleksen is a laminate of five fabric layers, which are arranged to form a resistive touchpad. The sensor is contact activated; when touched the layers are compressed together to form an electronic circuit.

Sensor- Textronics have developed a family of soft textile sensors used in wearable vital signs monitoring applications to sense the electrical activity of the body such as heart rate; or the mechanical movements of the body such as respiration.

Smart fabrics are an emerging technology first conceived in the early 1990s. Over the past 20 years scientists, technologists and manufacturers have developed them into viable market ready solutions, though there are still a number of technological challenges that will need resolving before they can become ubiquitous. These include connectivity and connectors, microelectronics encapsulation, power supply and storage and washability and durability [3].

13.3 ElekTex Fabrications Case Study

The most significant soft product using smart fabric that has made it to market, where the fabric has dramatically changed the functionality of the object is the Logitech KeyCase keyboard, which can be rolled up. Launched in 2002, it was enabled by ElekTex fabric developed by Eleksen. Beyond the novel rolling facility though, the keyboard has all the usual functionality of a keyboard and is styled to have the aesthetic language of a 'gadget', which can be connected to other technology driven mobile devices.

The products that are currently commercially available using the ElekTex technology can be starkly contrasted with the original range that was developed in 1999 by the design consultancy IDEO for the launch of the technology [4]. The brief for this project entitled 'Fabrications' was speculative with the ambition of promoting the possibilities of the technology rather than developing market ready solutions. Fabrications aimed to promote soft technologies as having the "potential to alter the use, appearance, tactility, function and cost of many products, across a range of industries" [5]. During the project they developed a range including a soft wrist phone, soft phone, conference phone, soft keyboard and soft remote control.

The Fabrications collection began to explore the materiality of fabrics used within the realm of product design but as with most commercially driven projects the focus was on developing outcomes. No time was spent discussing or developing an appropriate methodology to support the design of these essentially hybrid objects. The products designed during the Fabrications project were done so by an industrial designer and prototyped in a model-making workshop, they were not conceived at a cutting table or prototyped by a pattern cutter. Their point of reference lies firmly in the world of product design. In a case study of the project [6] one of the key points that emerged after interviewing designers who worked on the brief was that product designers need to begin to understand the cultural differences associated with using textile materials.

13.4 Cultural Values

Product design has cultural roots that are "linked to the development of industrialisation and mechanisation that began with the Industrial Revolution" [7]. Due to the often-long development cycles and associated investment to bring a product to market it has developed a systematic way of working that aims to minimise and control risk [8]. Product design also often professes to make people's lives better through the designs it produces, for example according to the Design Council, the discipline is described as "making things better–better for consumers and users, better for business, and better for the world" [9]. Might this set of values and rationalised approach to development not be as appropriate when designing soft products?

Fashion has its roots in the courts of late seventeenth century Paris [10]. There "lies in fashion an element that modernity would not have wished to acknowledge. Fashion is irrational. It consists of change for the sake of change, whereas the self-image of modernity consisted in there being a change that led towards increasingly rational self-determination" [11]. Fabric's often close proximity and association with the body "carries enormous social, cultural, political and moral weight. It is closely bound up with individual anxieties and broader social and historical concerns about the regulation of bodies in social space" [12]. Fashion is also often comfortable with being an appropriate "arena in which to investigate the complexities of modern life" [13]. When designing soft products should product designers be concerned with and inspired by issues that are important to the world of fashion such as "a fascination with image and narrative" [14] as much as human focused solutions that centre around functionality?

13.5 Engaging Product Designers

To begin to explore some of these ideas a class of third year product design students, from the University of Dundee, were asked to work with textile materials over the course of a 12 week semester [15]. Students were introduced to working with textile materials through a series of hands-on workshops, which included sewing and pattern cutting. These were then followed by workshops, which introduced a range of technologies that could be used with textiles such as laser cutting and the use of smart textiles using Arduino LilyPad technology. Students were also exposed to other aspects of fashion thinking by informal presentations in the studio and subsequently asked to embody their newly acquired knowledge in the design and make of a simple soft object.

The aim of this study was to begin to explore how a more meaningful engagement with smart textiles and the development of new soft products may be developed in the future for product designers. Contemporary fashion designers were introduced throughout the module and often alongside the work of product designers. For example, when looking at construction details using fabric, we looked at the intricate folding methods used by Issey Miyake as well as the inflatable products and temporary architecture created by the company Inflate. When looking at the use of smart fabrics and electronics in textiles, we explored the experimental fashion of Hussein Chalayan as well as the IDEO Fabrications project. By observing the students' design processes and outputs it was possible to begin to suggest how a dissolution of boundaries between the disciplines of product and fashion design might be catalyzed. The findings of this study placed emphasis on where areas of fashion design could be incorporated into product design.

Materiality- One of the key requirements to using textiles effectively is understanding and exploiting their material qualities. They are flat but can be turned into something 3-dimensional. They are soft and can be handled in ways that other materials cannot. They drape, they stretch and they move. To develop soft products the materiality of textiles needs to be understood by product designers.

Prototyping- One of the liberating features of working with textiles for product designers is that prototyping can be very quick and immediate compared to traditional hard materials. You can also prototype in a material very close to the one you will be finally using. Students were encouraged to create 'toiles' using calico to develop the shape through pattern cutting. There is certainly still some forward planning required but this way of working ultimately opens up a new perspective on prototyping for product design and should be fully appreciated.

The students' perspectives on their experience of working with textiles were sought through semi-structured interviews. One of the most illuminating discussions was over the perceived barriers to using textiles in product design. For many students it was beyond their preconceptions to use these materials before the module, one even commenting that she had not felt that she was "allowed to do anything like that." Many product design students mentioned that they think more in terms of solid objects than soft objects. Most students' last experience of working with fabrics was in their early years of secondary school during home economics classes. Perhaps it is no wonder that so few products are conceived as soft products.

13.6 Visions for the Future

The vision for the Eleksen/IDEO Fabrications project is inspiring. The designers developed a range of electronic consumables that really tried to exploit the potential qualities of smart fabrics. The success of the project was underpinned by a clear understanding of the notion of materiality and began to explore how textiles could impact on the world of product design. This paper argues though, that as the emerging field of smart fabrics matures the thinking developed during the Fabrications project could be extended even further to include new methods for designing soft products. Smart fabrics have the potential to revolutionize the way we think about and design some products but this vision will not be achieved unless product designers reconsider some of the cultural values that exist around their discipline.

Fabrics are traditionally the material used by fashion designers who have an intimate understanding, developed over centuries, of how to work with them. This paper proposes that the potential of soft products will only truly be realized if product designers embrace the world of fashion design, on some level. This will enable new ways of thinking and working to evolve that truly exploit the material qualities of the emerging technology.

References

1. Wilson A (2011) The future of smart fabrics. Smart fabrics 2011, London, 4–6 April 2011
2. Reho A (2011) Textile sensing in sports and medical areas. Smart Fabrics 2011, ditto
3. Dalsgard C (2011) OHMATEX subsequent exploitation of publicly funded R&D. Smart Fabrics 2011, London, 4–6 April 2011
4. IDEO (2011) Fabrications for Eleksen. http://www.ideo.com/work/fabrications. Accessed 25 May 2011
5. Chapman et al (2001) Fabrications. Electrotextiles, London
6. Duplock P (2005) Hybrid working practices between fashion and product design. Avantex Symposium, Frankfurt, 6–8 May 2005
7. Heskett J (1980) Industrial design. Thames and Hudson, London
8. Baxter M (1995) Product design. Chapman & Hall, London
9. Design Council (2011) Product design: the essentials of product design. http://www.designcouncil.org.uk/en/About-Design/Design-Disciplines/Product-design/. Accessed 13 Jan 2011
10. Breward C (2003) Fashion. Oxford University Press, Oxford
11. Svendsen L (2006) Fashion: a philosophy. Reaktion Books Ltd, London
12. Entwistle J (2003) The dressed body. In: Welters L, Lillethun A (eds) The fashion reader. Berg, New York
13. Evans C (2003) Fashion at the edge. In: Welters L, Lillethun A (eds) The fashion reader. Berg, New York
14. Wilcox C (2001) Radical fashion. V&A Publications, London
15. Duplock (2011) Soft product couture. In: 9th international European academy of design conference. http://endlessend.up.pt/files/EAD9_programabstracts.pdf. Accessed 16 May 2011

Chapter 14
Autonomous Mobility Scooter as an Assistive Outdoor Tool for the Elderly

Anthony Ntaki, Ahmad Lotfi and Caroline Langensiepen

Abstract This paper presents a conceptual design for a mobility scooter capable of outdoor autonomous navigation in an urban environment. The presented design of the mobility scooter has three stages. The first being the modification of a mobility scooter's current drive system to accommodate steering by wire on top of manual steering. The next stage presented is the design of a suitable on-board sensor placement configuration. This design configuration aims to place the navigational sensors on the mobility scooter in locations that are neither obstructive to the passenger's vision and movement while still maintaining a free and clear field of sensing. The paper presents a configuration that allows the placement of the sensors not to alter the original outlook of the mobility scooter. The third and final stage of the design is the development of the autonomous navigation system of the mobility scooter. The paper presents an overview of an outdoor navigational algorithm that fuses GPS and laser sensor data to follow a route in an urban environment while avoiding obstacles.

Keywords GPS sensors · Autonomous navigation · Assistive device

14.1 Introduction

A mobility scooter is defined as a mobility aid for those that have difficulty with walking long distances. It is powered by an electric motor and is steered with a handle bar that is attached to the front wheels. Mobility scooters come with four or

A. Ntaki (✉) · A. Lotfi · C. Langensiepen
School of Science and Technology,
Nottingham Trent University, Nottingham, NG11 8NS, UK
e-mail: anthony.ntaki@ntu.ac.uk

P. Breedon (ed.), *Smart Design*,
DOI: 10.1007/978-1-4471-2975-2_14,

three wheels. The back has two wheels while the front can have one or two wheels. An electric DC motor is connected to the back wheels and this is how the mobility scooter achieves motion. The front wheels are responsible for steering. They are connected to a handle bar that is controlled by the user of the scooter. The user controls the steering of the mobility scooter by applying a rotational force to the handle bar. The width and diameter of the wheels corresponds to the power provided by the electric DC motor of the scooter. A higher powered motor allows for wider wheels which in turn allow the mobility scooter to travel over rougher surfaces with a moderate speed.

When it comes to the legislation involving the classification of the mobility scooter as a member of traffic, a few things stick out. The first is that due to the fact that the scooter has not been around for a long enough period, many countries are finding it hard to find an appropriate classification for the mobility scooter [1]. Most countries classify the scooter within the same category as a wheelchair [2]. The problem with this is that while the wheelchair is also a mobility aid, some of the privileges offered to wheelchair users do not apply to mobility scooter owners. These privileges include reserved spaces on public transportation like buses and trains [2]. Other countries simply do not have a road network that is suitable for mobility scooter use. In such cases, the use of a mobility scooter as a mode of transportation has been restricted to private property [2]. In the United Kingdom, mobility scooters are classified as invalid carriages that are split into three classes [1]. These classes state the speed limitations of the devices and consequently the environment in which they can be used.

Mobility scooters are a common sight on pedestrian walkways nowadays. They are used mainly by the elderly to go shopping and visit their friends and relatives. The scooters are popular because they are easy to operate and one is not legally required to hold a license to use it. Mobility scooters are classified as invalid carriages. This means that they are driven on the pavements and pedestrian walkways. The mobility scooter is an assistive device. It is recommended for individuals that have difficulty walking over distances. The user of the mobility scooter needs to have two characteristics in order to operate a mobility scooter. First, the user needs to have the mental capacity to operate a scooter. This includes knowledge about what the buttons on the control interface do and how to read the gauges on the front panel. The user also needs to be competent in outdoor navigation. They need to know where they are and where they are going at all times during navigation. The second characteristic is upper body strength. The mobility scooter requires manual steering from the user. This means that if the user is to navigate the scooter, they need to possess the right amount of body strength to affect steering.

These factors have narrowed down the number and category of users that can benefit from the mobility scooter. With the modifications suggested in this paper, the mobility scooter can be used by a wider group of people. This will also benefit those that are already capable of using the scooter. The modifications will provide an extra layer of safety and comfort. They will achieve this by not only providing navigational decisions but by also providing steering power.

The layout of the paper is as follows. The related work is discussed in the next section, and then the methodology is introduced next. This methodology involves three parts that show how the mobility scooter is modified into a vehicle capable of autonomous navigation. A look at the work in progress and then some conclusions and future work are presented last.

14.2 Related Work

Automation of the driver functions in road transportation was developed to enhance the comfort and safety of the driver [3–5] and other road users. This was achieved by providing information about the vehicle [6] and it's surrounding. This information included current location, heading and speed. The information also included the state of other road users. An example could be whether another user was close or on a collision course. Another way was to alert the driver when the system detected an unsafe command from the driver [3]. Some systems would have a protocol that would reject commands from the driver if they determined them unsafe. They were to improve the comfort of the driver. This was seen with the introduction of power steering that required minimal effort from the driver to turn the steering wheel. The same was also introduced for the foot pedals. The idea was to have the driver use less effort while driving.

These systems provided the user with information about their vehicle and surrounding. The systems would in some cases assist in applying the commands of the driver. The one thing that the system did not do is take over complete control of the vehicle. This function was always left to the driver. Some have believed that the system would have been able to take over complete control but was not implemented due to the liability implications. It is important to remember that the driver needs to remain within the control loop. This is because the driver is supposed to be more competent than the system.

As sensor technology improved, the idea of a fully autonomous vehicle was investigated. The aim of this research was to produce a vehicle that was capable of autonomous navigation. The main advantage of an autonomous vehicle being safer travel at higher speeds [7]. Some working models [8–11] have been demonstrated in real world environments under controlled situations. In each of these cases, the vehicles' drive system has had to be converted to a drive-by-wire system. In [8], the steering of the vehicle is handled by DC servomotors that are controlled by a computer through a PIC. The results have shown that though a fully autonomous vehicle is yet to be developed it is however possible to implement a navigational system that automates some of the driver's tasks through partial or complete control of the vehicle. This concept is applied to a mobility scooter so that it may provide navigational assistance to the user.

14.3 Methodology

This paper presents a concept design for an autonomous mobility scooter. This is done by outlining the modifications that are necessary for the integration of an autonomous navigational system developed in software. The modifications are categorised into 3 parts. Mobility scooters have the option of disassembly. This option is provided by the manufacturer to allow for convenient transportation of the mobility scooter. But this feature allows us to access most of the scooter's internal components. These components can then be altered or modified and the scooter can then be reassembled without requiring any heavy duty mechanical work like welding.

14.3.1 Movement Control

14.3.1.1 Steering

Conventional steering mechanisms in vehicles are mechanical systems that transfer force from the steering wheel to the road wheels of the vehicle. Due to the mechanical nature of this system, it is also able to provide feedback forces from the road wheels to the steering wheel of the vehicle. These feedback forces allow the driver of the vehicle to judge the nature of the road surface which in turn allows him/her to apply the correct steering force to maintain control of the vehicle. When it comes to implementing a steer-by-wire system, work has been devoted to correctly reproducing these forces [12, 13] since the steering wheel of the vehicle is no longer mechanically coupled to the road wheels. This project is using the steer-by-wire system to allow the autonomous navigation system to generate steering control signals for the scooter while maintaining the original mechanical design of the steering system. A closed loop control system for the steer-by-wire configuration will be achieved by having the motor controller monitor the steering angle of the steering motor to ensure that the control signal has been received and carried out. The on-board sensors could also be used to help fulfill this task. The aim of this modification is to give the mobility scooter the ability of steer-by-wire. The concept of this is to have a stepper motor connected to the steering rod of the scooter as shown in Fig. 14.1. When activated, the stepper motor should steer the front wheels as a user would have. A stepper motor is used as opposed to a DC or servo because it possesses both the power (torque) and precision required to actuate the steering rod of the mobility scooter. This modification also requires that the steering motor installed accept digital signals. This is because all the navigation and drive signals are generated by the central processor that happens to be a laptop computer. Computers are capable of generating only digital output signals which is why direct connection with the stepper motor is not possible. It is possible to purchase a stepper motor that contains a circuit capable of

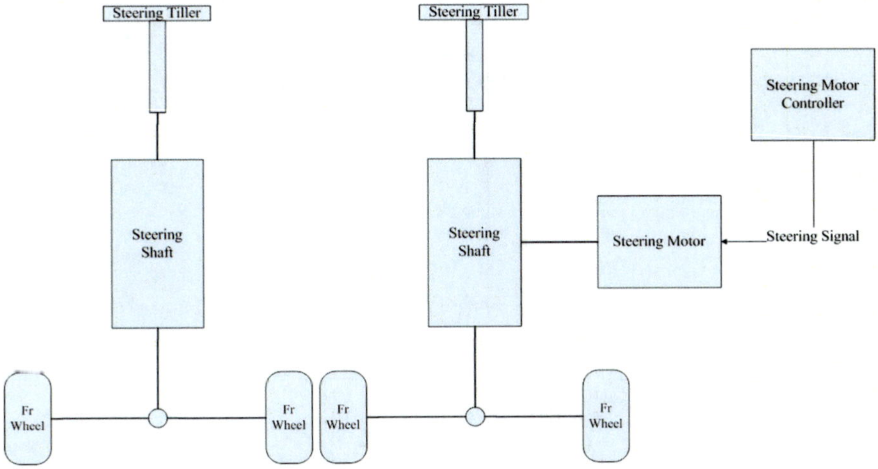

Fig. 14.1 Schematic diagram showing modification required to achieve proposed steer-by-wire system

converting digital signals into motor drive signals but in case this is not possible, there exist several PCB circuits on the market that can perform this function.

14.3.1.2 Drive Control

This modification aims to let the user send drive commands to the back wheel motor over a laptop computer. This will demonstrate the ability of the computer to generate drive commands like forward, reverse, slow and fast. The aim is to use as much of the existing components as possible thereby ensuring that not a lot of changes are made to the original vehicle.

This modification leads to the development of two components. The first component's task is to get input from the user and then convert and pass this input on to the central processor. This acts as a user interface. The user generated signals can be transferred directly to the central processor or they can be converted first by the component's on-board circuitry into processor friendly signals before transmission to the central processor. This conversion mainly involves analogue to digital signal conversion. The principal is that if the interface allows the user to input analogue drive signals, these will have to be converted to digital signals before they can be processed. This is why it is favourable to use an interface that only allows the user to input digital signals. However, some instances require analogue input for effective driving and navigation. These instances are investigated further on in the paper.

The second component has the task of converting the digital signals from the processor into ones that move the mobility scooter's actuators. These signals from the central processor are sent to the motors that move the back wheels of the scooter.

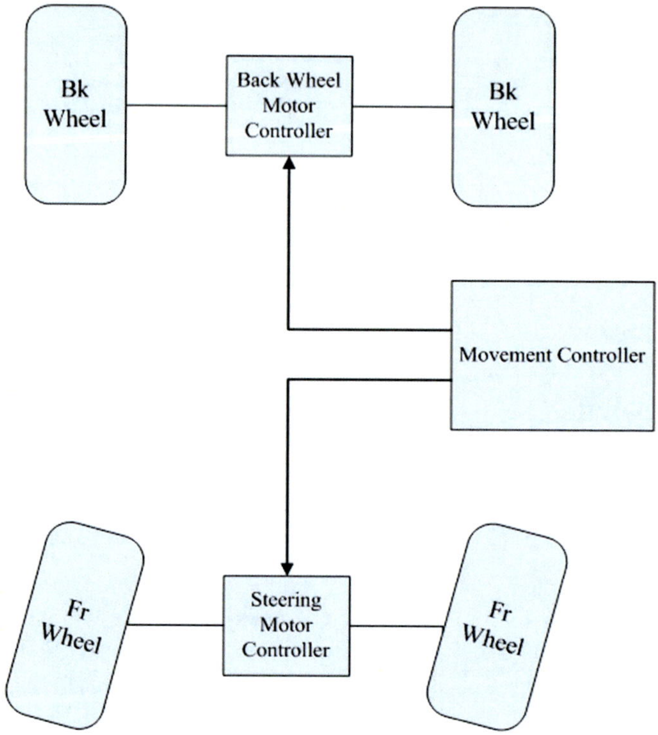

Fig. 14.2 Schematic diagram showing movement controller architecture

The mobility scooter already has a'motor drive unit' which is a collection of circuits that are designed to convert digital signals into drive signals for the motors. This means that this component is only required to send the correct digital signals from the central processor to this motor drive unit to achieve the desired effect.

The integration of the drive and steering controllers allows the mobility scooter to be fully controlled by an electronic system. This designed movement system then forms the basis for the movement controller as shown in Fig. 14.2.

14.3.2 Sensors

The third modification is to do with the placement of the navigational sensors. These sensors have to be placed in such a way that they achieve the following factors:

- They need to have sufficient access to the environment. The sensors should have their active sensing part unobstructed by the body and chassis of the mobility scooter

- The sensors need to be enclosed by the body of the scooter as much as possible. Referring to the point above, the sensor should have its inactive sensing part inside the body of the scooter and only allow the active part free to carry out its function.
- The sensors should in no way obstruct the movement of the user of the scooter.

The points presented above provide a guideline on how the sensors are placed onto the scooter. A typical navigational sensor is characterised by two parts. These parts are the main body and the connection from the sensor to the laptop computer. It is these parts that dictate how convenient it is to locate the sensors on-board the scooter. The main body of the sensor is placed in such a way that allows it to measure the specific aspect of the environment that it is meant for while keeping it securely fastened to the scooter's body. The connection between the sensor and the central processor on the other hand can be covered from view. This means that a space within the scooters interior can be used to house the connecting cable in a way that not only keeps away of the user's field of movement but also keeps it securely fastened. This minimises the risk of the connection cable being dislodged by uneven and sudden movement of the mobility scooter consequently avoiding navigational errors brought on by invalid sensory data.

The factors mentioned above also help provide some subtlety to the on-board sensors. This serves two purposes. The first is to try and maintain the scooter's current weight and dimensions. This is important because the mobility scooter's classification as a vehicle that can access pedestrian sidewalks is based on its weight and dimensions. Altering these specifications could lead to its reclassification as a vehicle unfit for movement on pavements which is where the navigational algorithm assumes is the main navigational environment. The second purpose is to preserve as much of the mobility scooter's visual look as possible. The assumption in this instance is that the current look of the mobility scooter represents the best possible aesthetic look and the sensor placement is supposed to maintain this as best as it can.

14.3.3 Autonomous Navigation

The modifications aim to make a common laptop computer the central processor of the mobility scooter as shown in Fig. 14.3. This means that all navigation related input and output passes through the central processor of the mobility scooter. This will then allow the processor to not only monitor all navigational commands but to also generate its own.

As the previous steps have transformed the mobility scooter into a vehicle capable of sensing the environment and accepting electronic drive signals, the final stage sees the design of a system that generates navigation decisions based on the data collected by the on-board sensors. It is at this stage that the autonomous navigation system is designed. There are two common architectures that are used

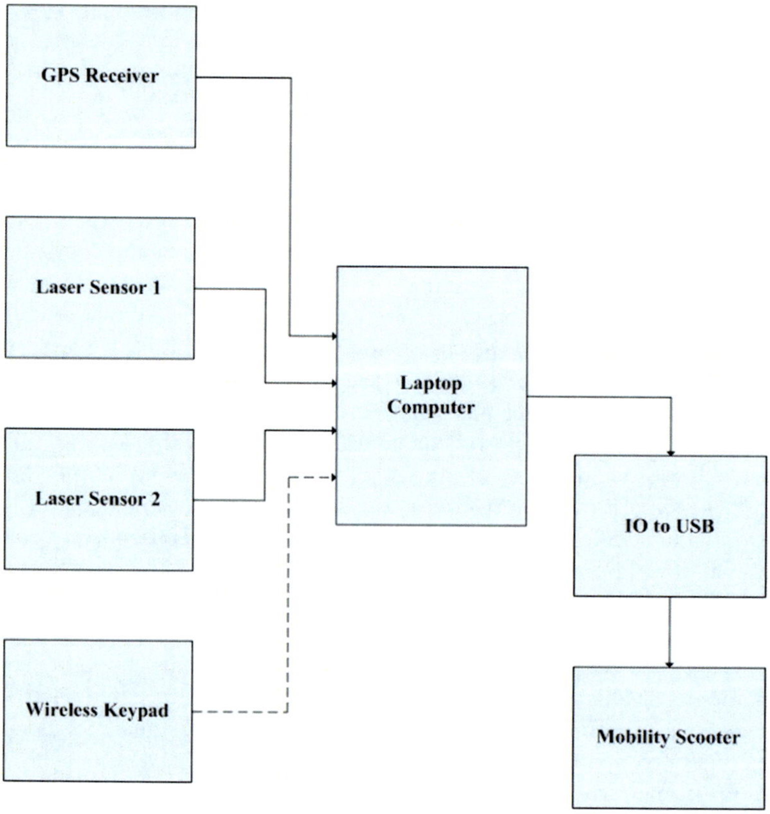

Fig. 14.3 System architecture showing components' connectivity

to achieve autonomous navigation and they include modular and hierarchical. Both architectures are meant to provide clarity to the system but they propose different ways of achieving this. Modular architecture emphasises simplicity by arranging the components of the system into modules with each module capable of independent functionality. Hierarchical architecture proposes simplicity by building components on top of other components.

Using the modular system architecture, the autonomous navigation system was designed. The navigation system comprises of behaviours that are executed according to a specified subsumption criteria. These behaviours are each designed to help the system navigate to a specified goal in a real world outdoor environment. Each behaviour has an input that is connected to a sensor and an output that is connected to the movement controller. Since there are multiple sensors and only one movement controller, we have multiple behaviours all sending messages to the movement controller simultaneously. To help with the behaviour selection, a

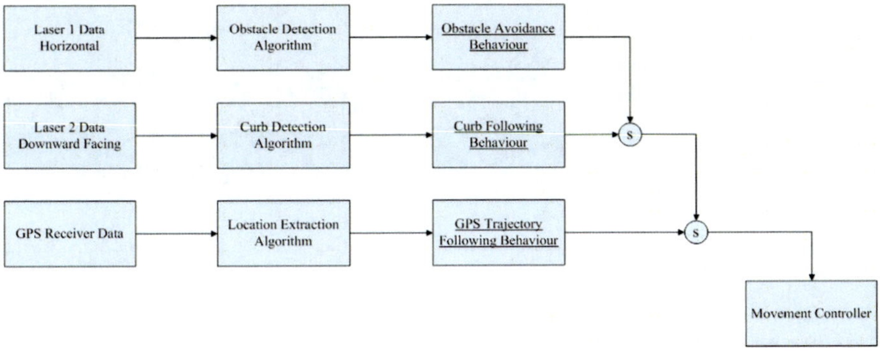

Fig. 14.4 Subsumption architecture of navigation behaviours

subsumption architecture is employed as shown in Fig. 14.4. In this particular architecture, when there is a connection, the higher behaviour takes priority.

The behaviours need to achieve three functions. These functions are obstacle proximity detection, environment feature detection and localisation. The obstacle detection behaviour detects objects that lie within the path of the scooter with the intention of avoiding collisions with these objects. A laser sensor is capable of performing this function within a specified sensor radius. Environment feature detection involves identifying a specific element in the environment that is crucial to navigation. In an urban environment, this feature could be a road, sidewalk, or pavement. Unlike the obstacle detection behaviour that avoids the detected feature, this behaviour follows the detected feature the same way a car follows a road to get to its destination. Given the fact that a mobility scooter navigates on a sidewalk, a curb is determined to be one of the environment features that is suited for this task. A curb is a common feature in an urban environment and it functions as a partition between the road and the sidewalk. A laser sensor that is mounted in a downward facing position has the ability to detect a curb by identifying differences in surface height with minimal detectable difference determined by the sensor resolution. The third localisation behaviour is responsible for determining the scooter's position on the surface of the earth. This information is also used to calculate the scooter's directional heading with respect to its goal or destination. A GPS receiver is used to achieve this function. These three behaviours are combined to achieve navigation in an urban environment. The combination of the behaviours helps address some of the limitations of the individual behaviours. The localisation behaviour is accurate to 3 metres but this is supplemented by the environmental feature detection behavior that provides more accurate information about the surrounding. From this it can be seen that the environment feature behaviour is able to determine the state of the local environment accurately but unable to provide a global view of the environment while the reverse is true for the localisation behaviour.

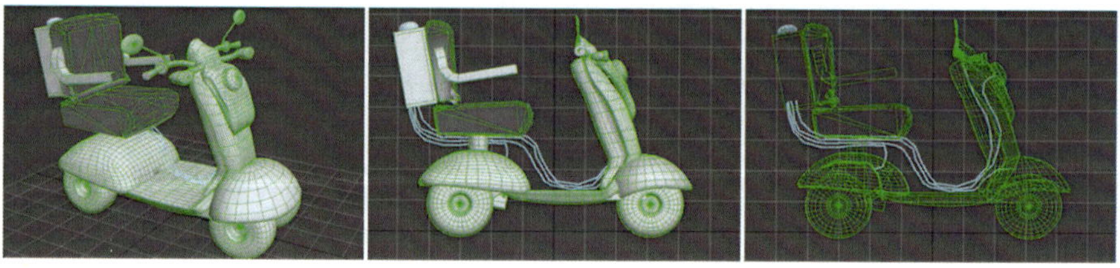

Fig. 14.5 Concept design for a mobility scooter showing the processing unit at the back of the seat

14.3.4 Conclusion

This paper offered a conceptual design for a mobility scooter as shown in Fig. 14. 5 that is capable of autonomous navigation. The design tries to maintain the original outlook of the scooter allowing it to retain its original aesthetic features. The modifications have shown that by digitising the movement control of the scooter, a navigation system can be implemented. This system could be used to provide navigation assistance by providing useful information to the user or it could take over complete control of the mobility scooter.

The next step is to realise the design using laser and GPS sensors connected to a laptop computer running the proposed behavioural navigational algorithm.

References

1. Transport, Travel, Research, and Ltd. Review of class 2 and class 3 powered wheelchairs and powered scooters (invalid carriages). Technical report, Department for Transport, 2006
2. UK Department of Transport (2006) Carriage of mobility scooters on public transport—feasibility study. Technical report, Department for Transport
3. de la Mara Armingol J, Escalera A, Hilario C, Collado J, Carrasco JP, Javier Flores M, Manuel Pastor J, Jose Rodrguez Ivvi F (2007) Intelligent vehicle based on visual information. Rob Autonomous Syst 55(12):904–916
4. Haynes NA, Moore WR, Martin AG (2008) Reliable microprocessor control for drive-by-wire car steering for the disabled. Microprocess Microsyst 6(7):361–366 9 1982
5. Stanton NA, Young M, McCaulder B (1997) Drive-by-wire: the case of driver workload and reclaiming control with adaptive cruise control. Saf Sci 27(2–3):149–159 12
6. Dorissen HT, Drkopp K (2003) Mechatronics and drive-by-wire systems advanced non-contacting position sensors. Control Eng Pract 11(2):191–197 2
7. Siciliano B, Khatib O (2007) Springer handbook of robotics. Springer, Secaucus
8. Bohren J, FooteT, Keller J, Kushleyev A, Lee D, Stewart A, Vernaza P, Derenick J, Spletzer J, Satterfield B, (2009) Little ben: the ben franklin racing team's entry in the 2007 DARPA Urban challenge, The DARPA Urban Challenge, pp 231–255
9. Montemerlo M (2008) Junior: the Stanford entry in the urban challenge. J Field Robot 25:569–597
10. Rauskolb F (2008) Caroline: an autonomously driving vehicle for urban environments. J Field Robot 25(9):674–724

11. Thrun S (2006) Stanley: the robot that won the darpa grand challenge: Research articles. J Robot Syst 23(9):661–692
12. Yixin Yao and Visteon Corporation (2006) Sae technical paper series 2006-01-1175 vehicle steer-by-wire system control
13. Yih P, Gerdes JC (2005) Modification of vehicle handling characteristics via steer-by-wire. Control Syst Technol IEEE Trans 13(6):965–976

Chapter 15
Emotional Design: Understanding Immaterial Values for the Development of Products

Maura Della Flora Flores and Liane Roldo

Abstract Professional design is a much broader area than simply designing products to fulfill the needs of individuals. According to Sudjic [1], the design goes beyond the obvious meaning of function and purpose of the objects. It is a language used to create objects that can reflect emotional and cultural values. This way, it is believed that industrial companies understand immaterial values inherent to the objects that can be produced so that they can be extended to new markets; these values can be understood through the knowledge of emotional expectations the individuals have about their products. In this context, this article is a experiment that uses some existing methodologies to understand the behavior of a particular group of people, the objects that have the greatest appeal to them, and the strongest connection with their cultural context.

Keywords Product development · Industrial design · Immaterial values · Emotional design · Quantitative research · Qualitative research · Memory · Material culture · Intangible factors · Conceptual projects

M. D. F. Flores (✉)
Campus Universitário da Região dos Vinhedos. Centro de Ciências Exatas, da Natureza e Tecnologias. Curso de Design de Produtos, Universidade de Caxias do Sul, Caxias do Sul, Brazil
e-mail: mauradf@gmail.com

L. Roldo
Departamento de Materiais. Laboratório de Design e Seleção de Materiais, Universidade Federal do Rio Grande do Sul, Porto Alegre, Brazil
e-mail: iroldo@uol.com.br

P. Breedon (ed.), *Smart Design*,
DOI: 10.1007/978-1-4471-2975-2_15,

15.1 Introduction

Several factors should be considered in the development of industrial products, from their adequacy to a particular industrial area to the meaning that a product may have on the lives of its users. This happens because when the objects are used there is an interaction between them and the person that handles them, that is, this relation generates an experience.

In design, when we refer to artifacts, the idea is that they should be able to fulfill the functions for which they were designed. In addition, they can also represent something more to their users. The objects are means used to promote experiences and to establish relationships with the individuals who use them.

The objects reveal much about their users. As Sudjic described, they are also the way to measure our life transition. He added:

> Objects are how we define ourselves, to signal who we are, and what we are not, whether jewellery, furniture or clothing [1].

Through these principles, this study was carried out to understand how individuals relate to some of their objects and what type of value they attribute to them. The qualitative and quantitative methods were used for the development of this study.

The qualitative research consisted of a survey with individuals born in the State of Rio Grande do Sul; in southern Brazil (these individuals are called *Gaúchos*). It aimed at questioning them about the objects that they refer to in this geographical region. The categorization of their responses was based on Mihaly Csikszentmihalyi and Eugene Rochberg-Halton [2] survey that provided a singular perspective on the affective relationship of people with their common domestic objects. Also, this method also finds justification in Mirian Goldenberg that recommends the use of qualitative research to study the "issues that are difficult to quantify, such as feelings, motivations, beliefs and attitudes" [3].

On the other hand, the quantitative research referred to the studies by Lindstrom [4] that quantified the five senses in situations involving people's interaction with products. Based on these concepts, flow diagrams were developed to quantify features previously recognized and classified during the survey.

15.2 Relationship Subject X Object

As observed previously, the object always refers to certain a place, whatever it is. It is connected to the experience of the individuals with and in the world, since it represents a significant portion of the landscape where they live [5]. This way it is possible to associate the objects with something essential in the environment where people live mainly because they constitute "one of the basic references of the individual's contact with the other" [6]. In this context where the relations are established between the subject and objects, Damazio emphasizes that "it is inconceivable to imagine us today without the 'extensions' that we develop to meet our needs, from the simplest to the most complex, from the most basic to the most transcendental" [7].

This experience involved in the use of a product also "refers to the connection people have with places where they live, being it a natural and particular experience by the cultural expression where the objects are found" [8]. These objects rooted in society are at the same time "a product and a reflection of its cultural, political and economic history. Therefore, it helps to shape the society affecting the individuals' quality of life," as described Norman and Draper [9].

According to Russian and Moraes [10] people, in addition to using the products for their first functions, still look for something extra in them, as they are accustomed to the way they normally function; they wish for pleasant products which they can relate to. The same authors also suggest that to design nice products, one must understand the consumer completely, how people use the products, the role they play in their lives, and how the relationship between the person and the product occurs.

This way of using the objects, according to the role they play in the people's lives, characterizes different experiences and reveals the existence of a relationship with the subjects, and was summarized by Damazio when he stated that the artifacts:

> [...] participate actively in everyday life, and are used by people to establish meanings about themselves and their lives. Sometimes they concentrate social actions. When interrelated with other mechanisms, the artifacts can instruct individuals about the world organize and regulate social practices, influence behaviors and cause real and different effects (Damazio, 2005, p.49) [11].

She also describes that things that surround individuals are inseparable from what they are, in addition to being the basis on which the meaning their personal lives is built. Menezes [12] complements by saying that it is necessary to understand what happens between individuals and material things to understand what people are.

After mentioning these factors related to the relationship between people and the objects that surround them, it is possible to search for references in the field of anthropology. By joining anthropology to design, we seek to understand the new requirements described by Iida and Muhlenberg

> Rational thinking rejects solutions derived from intuition and emotion, considering them as second-order elaborations. However, recent developments in design place pleasure and excitement as the focus of a new methodology, taking them from the marginality to which were relegated [13].

In addition to objects being used by people, they are also hated or loved by them, characterizing the objects as artifacts that, besides possessing physical forms and mechanical functions, assume a "social form" and a "symbolic function" [14]. This encourages designers to change their way of seeing design. The author adds,

> Designers turned their attention to people and how they interpret and interact with the physical and social environment. Then, they begin to design with a focus on emotions and with the intention of providing pleasant experiences [15].

Emotional Design, also known as Affective Design, is being currently discussed because it is believed that the products created by designers aim at providing pleasurable experiences and establishing emotional relationships with their users [15].

The field of design and emotion can be recognized, also from terms such as Experience Design. Its practice strives to meet not only the immediate needs of users, "but also to understand and fulfill the human motivations for the product, being related to small and great life experiences" [16].

15.3 Design and Culture

By contextualizing design, culture and the relationship between them, this study initially seeks for references on the concept of culture; for Tomaz Tadeu Silva it is a field of meanings in which different social groups want to impose their meanings to a broader society. Here, culture is considered a place for the contraposition of meaning. What is involved in this situation is the definition of cultural and social identity of different groups [17].

To contribute with this definition, Ono argues that the concept of culture is fundamental for the analysis of cultural diversity. Edward Burnett Tylor, in 1871, provided an all-inclusive definition of culture as "that complex whole which includes knowledge, belief, art, morals, law, custom, and any other capabilities and habits acquired by man as a member of society" [6]. The author also points out that, generally, despite the different approaches, the conceptual approaches of culture include both "immaterial" (the habits and products of the mental activity, for example), as the material elements of human societies.

Culture, in this study, does not intend to discuss the different currents and existing cultural events; it is restricted to the study of culture, in its tangible and intangible elements of an identity established in specific geographical delimitation. Chaves words suggested this definition:

> The culture is not a "scope" or "field," but a dimension that permeates all levels of social life. Thus, in some discursive contexts, "culture" is used as a synonym for community [18].

This way, we understand that culture includes tangible and intangible factors in a particular social context. The industrial designer has a role defined in the context of these societies and in the "process of developing concepts and materials reflecting the needs and issues of cultural diversity of individuals and societies" based on the "understanding of the roles that objects play in people's lives within the socio-cultural concepts where they are inserted" [19].

As for the definitions of material culture, Nogueira refers to the collective material culture as something that became diverse over the centuries, "due to the adversities imposed by the environment to which different people had to adapt" [20] Ono [6] relating it to the design area, paraphrases Bonsiepe when he says that "the designer is both the subject and object of a cultural dynamics. He influences the material culture and is influenced by it" [21].

To complement the definitions of material culture, it is necessary to know that it is derives from the characteristics of material goods in a given society. This way, the cultural object is defined by Norberto Chaves when ensures that:

> All cultural, material or immaterial good is such because it can be shared by an entire community, i.e., it holds a meaning of social impact. And it is easy to understand that there can be no possible meaning without some degree of coding that allows it to be registered. To live, understand and enjoy culture, we put into action a complex series of paradigms or codes, that is, in culture there are rules: the culture itself is a huge and tangled system of rules [18].

As the characteristics of a people do not become real just by considering what is touched, measured, the relevant analysis on such questions concerning the use of a product can be clearly understood. Such characteristics, which are also named as "intangible," can be defined, according to Karana et al. [22], as something that cannot be perceived by senses especially by touch, and are difficult to identify, for example: smells, tastes, memories, feelings, among others.

Immaterial characteristics are referred to in many ways. Karana et al. organized them as features of second and third order of materials [23], qualitative properties [24, 25], non-active or passive functions of materials [26], non-technical aspects of materials [27] material image, metaphysical aspects of materials, non-physical properties of materials [28], personality, personal dimension of materials [29, 30], intrinsic cultural meanings of materials [31], subjective, substantial dimension and material personality [32], perception characteristics of materials [33] and perception values of materials [34, 35].

15.4 Research Methods

Qualitative and quantitative approaches were considered in this study; in the first, scientists work with interpretations of emotions, values and subjectivity; in the second, scientists deal with external objects which may be known objectively [34].

15.4.1 Qualitative Research

In qualitative research, according to Goldenberg [3], the main concern is to provide a better understanding of a social group, an organization, institution, a trajectory, etc.

Some authors did some relevant researches that were used as references in this study and listed below:

"The Meaning of Things: Domestic Symbols and the Self"

Csikszentmihalyi and Rochberg-Halton carried out a study interviewing 82 family members from Chicago, in the United States, where they were asked what objects in their houses were special to them and why. The authors identified, in the houses they visited, the existence of a network of objects that gave meaning to their lives, but objects of art were rarely part of this network. Many of them had been made by children, relatives, or friends and these ties would be the value that gathered their owners and protagonists, not their visual qualities. Sculptures, paintings and drawings were dear to them because they reminded them of a place, a person or a special event.

With this study, the authors concluded that it was not what they denominated "quality of design" that made an object special and dear, but the type of interaction that was established between the person and object. The research data also suggested that an object does not create order in the mind of their observers because it embodies the principles of visual order, but because it helps them order their own experiences. As the authors described, people find meanings in objects that concretely embody the most important accomplishments, actions, attitudes and events of their lives.

The first part of this study was developed in the context of a qualitative research, with approximately 100 subjects being asked which object they thought had greater connection with their state (the geographic region where they live). The search for data about the objects of gaucho's cultural identity is given into stages: the first one contemplates a virtual space for acquisition of testimonials and the second one comprehended verbal discussions.

To develop a method for the testimonials, it was resorted to the studies of anthropologist Mirian Goldemberg were used to inform the research [3]. In her book, The art of research: how to do a qualitative research in social sciences, it was reported the sociology research practices in Chicago, made by researchers W. I. Thomas and F. Znaniecki, that studied the emigration of Dutch peasants. These scientists gathered the letters changed between those families and newspaper articles, to use as documentary source.

From the example given by him, the relation between acquisition means of data used in researches in the School of Chicago, it was elaborated a mean to acquire them. This research also collected testimonials from "documents and personal diaries," starting in the ways those articles are being read nowadays: from virtual space.

It was observed that a virtual tool could be used to acquire a relevant number of testimonials and also reach beyond people from the specific cultural region, Rio Grande do Sul, the ones that are leaving abroad, and others that had any interest to contribute to the research using the internet.

With the virtual site for the testimonial acquisition, it was created a Blog titled "Maura's Research".[1] The next challenge was to insert the visitors in the research

[1] www.pesquisadamaura.blogspot.com

context, becoming necessary to have a way of presentation so the participants would feel comfortable to tell their histories.

The Blog should, necessarily, have an initial text as introduction and instruction, with the focus to present the research to the visitors. Posterior to the text reading, the visitor would read the questions and in the section "comment," they could answer it.

The objective was to make the introduction text sound like an informal talk, as it follows:

> Hello, One of the instruments that I am using in my research is testimonials. The objective is to know the characteristics that constitute and reproduce the faces of the gaucho identity, as in Rio Grande do Sul, many cultural manifestations coexist. To recognize those characteristics, one of the methodological steps that I elected was to obtain testimonials and establish talks. This blog has the main objective to collect information from different points of views, about a few questions: Where were you born? If you are not a gaucho, since when do you live here? What object make you remember Rio Grande do Sul? If you posses this object, how did you get it? All comments are welcome!

Their responses provided a list of different types of products: physical (material) and immaterial "objects" as memories and smells, totalizing 75 testimonials.

Having some characteristics in common, these objects were classified following the references provided by Csikszentmihalyi and Rochberg-Halton [2] who mentioned that the respondents suggested their own classification as they were able to relate similar characteristics between different responses.

15.4.2 Quantitative Research

'Design for Emotion' Support via Human Sensations (Demo HS). The DemoHS is a project intended to guide designers to investigate the role the senses play in the user-product interaction. It is based on the theory that, after this interaction through senses occurs, sensations are generated [36] they result in the perception of stimulus after a comparison in the brain between the user's knowledge and experience. According to these authors, it is the evaluation of the user's perception of stimulus, his objectives, rules and attitudes that creates what stimulates emotion.

Preliminary tests for the DemoHS model submitted by Fenech in 2006 aimed at identifying the different stages presented in the model. Some attempts were made to prove the link between these phases. A series of products was presented to participants who were invited to interact with products for a short period. Following, interview sessions were carried out.

The number of senses and the degree to which a product appeals to the five senses can be identified and quantified with the use of a sensagram [37]. The sensagram is characterized by a flow diagram; in each sense vertex, a different sense is attributed. There is a scale that ranges from one (minimum) to five (maximum). The higher the rating, the more appealing is the product to the sense in question.

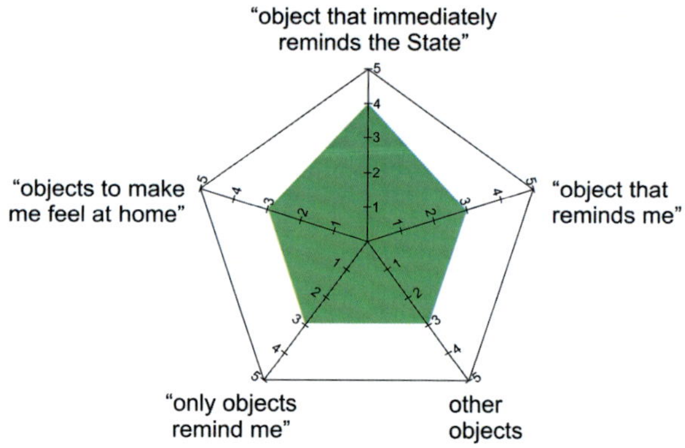

Fig. 15.1 Flow diagram of categories (adapted from Fenech and Borg 2006)

This sensagram provides a quantified sensory performance of a product. The products that most appeal to a specific sense cover the largest area. The shape shown by this diagram makes it possible to see if a product has strong connections to a particular direction; its score can even be processed to allow the comparison between different total values, proximity and distance from the sensory appeal of different products. According to Lindstorm'method, the qualitative results and their categorization were organized in a flux diagram using a quantitative diagram of five distinct categories, gathering four main categories and the other referred objects, called "other objects."

These five categories were organized quantitatively in the flow diagram below (Fig. 15.1).

In this flow diagram, four different categories, obtained from the interview responses, were listed as well as the other objects cited; the largest numeric value was assigned to the category, "the object that immediately reminds the State," as it was the most mentioned.

Further, the value 3 was attributed for the other categories, as they were referred to with the same frequency; this way, it was possible to view the responses obtained from the categories in a numerical and comparative way.

Even when each category is represented in the diagram, many of them have intrinsic characteristics with very important peculiarities that form detailed sub-categories that can useful in the development of projects. While analyzing these characteristics, the category "The object that immediately reminds the State" presented many aspects by which people identify the "*cuia*" and its peripherals as an object easily related to the State.

Among these relationships, the justifications were diverse; most of them revealed that the use of certain objects maintained the daily habits; traditions passed from father to son, and brought people closer together, among other factors.

Some objects reminded how objects emphasized the identity of the State because they were very important icons. These aspects were shown in the flow diagram, in Fig. 15.1. The values given to each attribute were in accordance to the frequency and relevance in which they appeared in the responses.

15.5 Conclusions and Proposals

With the results found in this study, some material and immaterial objects that represent the gaúcha culture were described. They are used daily, or are souvenirs for those who are away from the State of Rio Grande do Sul; they are the ones someone has received as a gift, those which are memories of childhood and are kept in our memory. Therefore, they were mentioned for the fact that they represent a situation, an encounter, a feeling of being close to home, through typical objects, popular sayings, smells, memories of landscapes and colors, habits, songs, photos, the *churrasco* and its aroma, sounds, wind, the State flag, the climate, the accent…

These relations also justify the connection of people with the objects previously mentioned, from relations of memory. According to Bosi [38] the permanence of the facts in one's memory depends on how affective was the impact.

The project of new industrial products that use these attributes as references mentioned by the people are able to bring up memories, and these, according to Norman make us bring back situations, reflecting experiences of our lives and things surrounding them such as family, friends, personal achievements and disappointments and also the way we see ourselves. These are the characteristics that define the associations we make, as Norman states: "what really matters is the history of interaction, associations that people have with the objects and the memories they evoke" [39].

The associations people are able to do with objects define our attachment to things, as Norman states describing the research of Mihaly Csikszentmihalyi and Eugene Rochberg-Halton, in The Meaning of Things:

> We become attached to things if they have a significant personal association, if they bring to mind pleasant, comforting moments. Perhaps more significant, however, is our attachment to places favorite corners of our homes, favorite locations and views: Our attachment is really not to the thing, it is to the relationship, to the meanings and feelings it represents [39].

This way, it can be also understood that not necessarily only three-dimensional objects evoke feelings and sensations in their users, but also non-tangible objects, from memories and memorabilia.

The cultural objects that are inherited are considered important like the gifts received from parents, which the sons/daughters will take with them when they move out of their original home. Together with the gifts, they also take with them the habit of taking "*chimarrão.*" This event can be evaluated as a way of preserving identity, from the cultivation of typical regional costumes.

In this context, a series of actions that refer to the State were also identified, primarily considered by people who have lived out of the State, such as "To eat bergamot in the sun," and the accent ("it was with me in every city I lived").

Therefore, what represents a people is not only identified with the material objects, but also with their meanings, the interrelations established through them, the memories that remind things, and the experiences that have been established.

The products generated because of the research may also establish social relationships, as typical objects from Rio Grande do Sul, considering that they represent something common for the *Gaúchos*, have different meanings for each individual as the result of their experiences. And as they have different meanings, there may be some exchanges when a person asks the other on the use of a certain product.

And yet, the use of products designed for individuals with common characteristics, can establish social relations when they are used outside their place of origin. They can also call the attention of people from different places. Their cultural reference sets out cultural exchanges and strengthens cultural identity, as these characteristics can be conveyed to other locations not only through the people, but also through the design products.

Both material and immaterial goods resulting from the methodological phase of this research were important subsidies for the creation of the conceptual design. Much of this understanding took a numerical form derived from a well-planned organization in the flow diagram, highlighting the data on the objects that had more participation in the interviews in relation to the others which had less.

This study has shown the relevance and, above all, the potential for transformation that the handling of the instrumental field of Design and Emotion can bring to the product design, and through knowledge of interests and meanings of a group, or social groups, through the use of a simple and efficient diagram, map and the establishment of parameters for the product design. Thereby, it has established the flow diagram as a technological tool. Besides:

- The interactive process by which several people were able to participate with their opinions generated a method to develop products;
- Based on the literature research in the area and data obtained in the survey, it is believed that the information described here can be used as a starting point to bring up issues related to emotional interaction with different types of users;
- Flow diagrams from the quantitative method were complementary to the qualitative research as they classify opinions, memories, experiences and different, intangible immaterial characteristics cited;
- The combination of these two methods provided a better understanding of the whole research, since this feature can be used as a methodology in Design and Emotion, as a technological contribution to the industrial processes;
- Regarding the industry, this methodological organization that brings together the quantitative method for the understanding of immeasurable features, even when they refer to the user's emotions and the feelings, can supplement the researches for trends for different individuals and expand the market segments;
- The use of flow diagrams can still be useful to understand the range of different products and also make comparisons between them by overlapping the results of the diagrams (as illustrated in Fig. 15.2), and the perceptive scope of each product can be compared considering the area of the diagram covered by it;
- Flow diagrams are also helpful in interpreting and identifying trends, needs and values active in the context it was applied, i.e., the project;

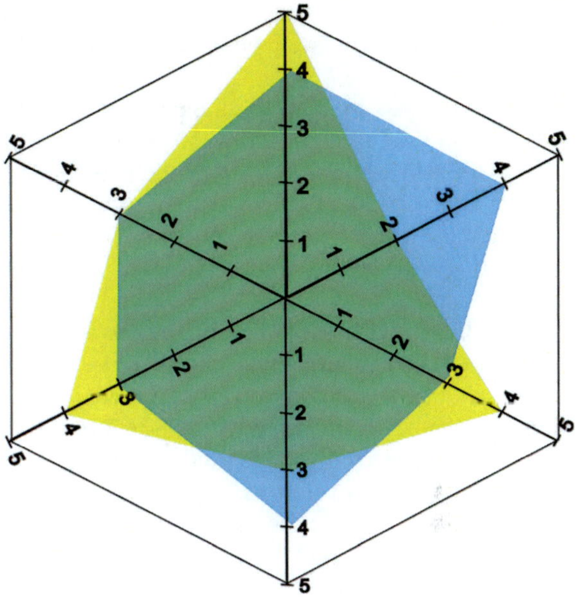

Fig. 15.2 Comparison between results from two different flow diagrams

- The contribution of this research also refers to the possibility of interacting quantitatively the qualitative elements in order to assist the designer with the project and to interpret the user's needs inserted in a given environment;
- Conceptual projects developed from this method are able to communicate aspects of the *gaúcha* identity in this case, the development of projects like these can improve the perception of people before their culture, allowing to know it better as well as their own cultural background and their own identity, in order to strengthen it;
- By studying ethnic design, it is possible to understand and consolidate regional identities; and from this, a possible identity for the *Gaúcho* and Brazilian design.

Firstly, the proposal is for the creation of a line of different products and souvenirs to be available commercially, representing the State in its various regions and other regions of the country. To create a line of new objects, four different categories obtained from the responses could still contemplate a greater number of features approached and ensure greater cultural influence.

Another suggestion would be to develop some researches in different regions, since that State presents many distinct cultural events and a comparative study could be conducted, and an analysis of the products that are relevant in each region.

Furthermore, it may be considered relevant to know cultural characteristics of other Brazilian states participating as the "foreign" places to carry out a comparative study of the most important objects for each of the cultures studied and compare their identities. This way, it would be possible to interpret in what way the products from a particular cultural identity could be inserted in another context, and how this adaptation would occur. Ethnodesign in Brazil, develops as a proposal capable of redeeming processes and technologies that are unique of Brazilian ethnic groups.

The author adds by stating, "the production of these groups contributes to the construction of the symbolic universe and identity in our country, as most of these products are part of the daily life of the population."

Another possible and relevant study would be to do research on what people carry with them when they move away from their country or place of origin, investigating what objects are elected as representatives of their culture, homeland, memories…

For all these studies, it is necessary, however, to recognize the values derived from the local and cultural environment, to be interpreted and decoded as possible attributes for the elaboration of industrial artifacts.

Finally, the development of this research reveals the possibility of interpreting the characteristics of local identities and ways to quantify them, enabling the development of the local tourist as well as contributing for the self-esteem of these populations.

References

1. Sudjic D (2010) A linguagem das coisas. Tradução de Adalgisa Campos da Silva. Intrínseca, Rio de Janeiro, p 21
2. Csikszentmihalyi M, Rochberg-Halton E (1991) The meaning of things. Domestic symbols and the self. Cambridge University Press, Cambridge
3. Goldenberg M (2005) A arte de pesquisar: como fazer pesquisa qualitativa em Ciências Sociais, 9th edn. Record, Rio de Janeiro, p 63
4. Lindstrom M (2005) Brand sense—how to build powerful brands through touch, taste, smell, sight and sound. Kogan Page, London
5. Silveira FLA da, Lima Filho MF (2005) Por uma antropologia do objeto documental: entre "A alma nas coisas" e a coisificação dos objetos. Horizontes Antropológicos, Porto Alegre, 11(23):37–50, jan/jun 2005
6. Ono MM (2006) Design e cultura: sintonia essencial. Edição da Autora, Curitiba, p 16
7. Damazio VMM (2005) Artefatos de memória da vida cotidiana: um olhar interdisciplinar sobre as coisas que fazem bem lembrar. Tese. (Doutorado em Ciências Sociais). Universidade do Estado do Rio de Janeiro, Rio de Janeiro
8. Silveira FLA da, Lima Filho MF (2005) Por uma antropologia do objeto documental: entre a "A alma nas coisas" e a coisificação dos objetos. Horizontes Antropológicos, Porto Alegre 11(23):37–50, jan/jun 2005, pp 40–41
9. Norman DA, Draper SW (1986) User centered System Design: new perspectives on human-computer interaction. Lawrence Erlbaum Associates, Londres, p 27
10. Russo B, Moraes A de (2005) Uma abordagem sobre a usabilidade de produtos prazerosos—um estudo de caso. Estudos em design, Rio de Janeiro
11. Damazio V (2005) Artefatos de memória da vida cotidiana: um olhar interdisciplinarsobre as coisas que fazem bem lembrar. Rio de Janeiro, 2005. 285. Tese (Doutorado em Ciências Sociais) Universidade do Estado do Rio de Janeiro
12. Menezes CS de, Damazio V (2007) Design e Emoção: Sobre a relação afetiva das pessoas com os objetos usados pela primeira vez. Dissertação (Mestrado em Artes e Design)—Pontifícia Universidade Católica do Rio de Janeiro, Rio de Janeiro

13. Iida I, Muhlenberg P (2006) O bom e o bonito em Design. In: Anais do 7° Congresso Brasileiro de Pesquisa e Desenvolvimento em Design. Curitiba, p 2
14. Mont'Alvão C, Damazio V (2008) Design, Ergonomia e Emoção. Rio de Janeiro: Mauad X: FAPERJ
15. Norman D (2004) Emotional design: why we love (or hate) everyday things. Basic Books, NY, p 11
16. Padovani S, Buccini M (2005) Uma introdução ao Design Experiencial. Estudos em Design. Rio de Janeiro, 13(2): 09–29, 2005, p 63
17. Silva TT (2000) Documentos de identidade: Uma introdução às teorias do currículo. Autêntica, Belo Horizonte
18. Chaves N (2006) El diseño invisible. Siete lecciones sobre la intervención culta en el habitat humano. Paidós, Buenos Aires, p 19
19. Ono MM (2006) Design e cultura: sintonia essencial. Edição da Autora, Curitiba, pp 16, 22
20. Nogueira JFS (2005) Etnodesign e Cultura Brasileira: memória, resgate e identidade. O Guardador de Inutensílios. v. 8, maio, pp 5–14, 2005. Disponível em: <www.neppi.org/anais/textos> Acesso em 17 out. 2008, 5
21. Bonsiepe G (2002) Entrevista concedida a Maristela Mitsuko Ono. Istambul
22. Karana E, Hekkert P, Kandachar P (2008) Material considerations in product design: a survey on crucial material aspects used by product designers. Em: Mater Des 29:1081–1089
23. Hodgson SNB, Harper JF (2004) Effective use of materials in the design process–more than a selection problem, International engineering and product design education conference, Tu Delft, 2–3 Sept 2004
24. Jee D-H, Kang K-J (2001) A method for optimal material selection aided with decision making theory. Mater Des 21:199–206
25. Edwards KL (2002) Towards more strategic product design for manufacture and assembly: priorities for concurrent engineering. Mater Des 23(7):651–656
26. Deng YM, Edwards KL (2005) The role of materials identification and selection in engineering design. Mater Des
27. Ferrante M, Santos SF, de Castro JFR (2000) Materials selection as an interdisciplinary-technical activity: basic methodology and case studies. Mater Res 3(2):1–9
28. Ljungberg LY, Edwards KL (2003) Design, materials selection and marketing of successful products. Mater Des 24(7):519–529
29. Ashby M, Johnson, K Materials and design: the art and science of material selection in product design, Butterworth-Heinemann, Oxford
30. Ashby MF (2005) Materials selection in mechanical design, 3rd edn. Butterworth-Heinemann, Oxford
31. Manzini E (1986) The material of invention. Arcadia srl, Milano
32. Rognoli V, Levi, M (2004) How, what and where is possible to learn design materials? International engineering and product design education conference, Delft 2–3 Sept 2004
33. Zuo H, Jones M, Hope T (2005) Material texture perception in product design, international conference on the art of plastics design, Berlin 18–19 Oct 2005
34. Karana E (2004a) The meaning of the material: a survey on the role of material in user's evaluation of a design object, 4th international conference on design and emotion
35. Karana E (2004b) Guidelines for a materials selection source for industrial design activity: a survey on the expectations of turkish designers Thesis (Ms) METU, Ankara
36. Fenech O, Borg J (2006) A model of human sensations as a basis for 'design for product-emotion' support. In: Proceedings of the Conference Design. Dubrovnik, Croatia
37. Lindstrom M (2005) Brand sense—how to build powerful brands through touch, taste, smell, sight and sound. Kogan Page, London
38. Bosi E (2006) O tempo vivo da memória: ensaios de Psicologia Social. São Paulo, Ateliê
39. Norman D (2004) Emotional design: why we love (or hate) everyday things. Basic Books, NY, p 48